Belief, Evidence, and the Unseen: Miracles, Evolution, and the Moon Landing

Ammar Adil

Published by Ammar Adil, 2025.

BELIEF, EVIDENCE, AND THE UNSEEN: MIRACLES, EVOLUTION, AND THE MOON LANDING

First edition. November 18, 2025.

ISBN: 979-8232587550

Written by Ammar Adil.

Also by Ammar Adil

Sunni Hadiths and Sirah: Proofs of Historical Reliability
The Existence of God and the Irrationality of Atheism
Proofs for Islam: Prophecies of Prophet Muhammad ﷺ
Hadiths Sunnites et Sirah : Preuves de Fiabilité Historique
Evolution and Design: Logic and Evidence
Evrim ve Tasarım: Mantık ve Kanıt
Sunnitische Hadithe und Sirah: Beweise historischer Zuverlässigkeit
Sünni Hadisler ve Siyer: Tarihsel Güvenilirliğin Delilleri
From Skepticism to Certainty: The Doubting Muslim's Guide
Évolution et Conception : Logique et Preuves
Evolution und Design: Logik und Beweise
Du scepticisme à la certitude : le guide du musulman qui doute
Vom Skeptizismus zur Gewissheit: Der Leitfaden für zweifelnde Muslime
Beyond Scientific Miracles: Islam, Science, and the Search for Truth
L'Existence de Dieu et l'irrationalité de l'athéisme
Tanrı'nın Varlığı ve Ateizmin Mantıksızlığı
Belief, Evidence, and the Unseen: Miracles, Evolution, and the Moon Landing

Table of Contents

Preface

This book has been a very long time in the making. In modern times—especially in the West, but increasingly throughout the world—belief in miracles is seen as irrational, rooted in desperation and blind faith. By contrast, evolution is widely accepted as fact. It is regarded as something proven beyond any reasonable doubt, supported by numerous lines of evidence. Anyone who rejects evolution in general, or human evolution in particular, is often dismissed as irrational or deluded.

A similar situation exists with the moon landing in 1969. It is widely accepted by Western academia, mainstream media, governments, and educational institutions around the world. Yet, in many interesting and obvious ways, it bears striking similarities to a miracle. To give just one example: the moon landings are said to have taken place decades ago, yet they are not happening today. One of the most common objections skeptics raise against miracles is precisely this—why do they no longer occur?

I have been researching the existence of God and other great questions for many years, engaging with and debating countless atheists and skeptics. After a long process of study and critical reflection, I came to the conclusion that the modern debate on God's existence is riddled with double standards and dishonesty—not only among atheists and skeptics, but even within mainstream Western academia.

The need for a fair standard of reasoning and a critical examination of atheist assumptions led me to write my earlier books: The Existence of God and the Irrationality of Atheism and Evolution and Design: Logic and Evidence. In those works, I examined the criteria and standards atheists and evolutionists use to justify their beliefs. I demonstrated that arguments for God often meet those very criteria, while the arguments against belief collapse when those same standards are consistently applied.

In this book, I aim to extend that same approach to the topic of miracles. I will explore the logic that underlies the debate about their possibility or impossibility, beginning with what qualifies as a miracle and addressing some issues of definition. I will also examine specific physical miracles attributed to Prophet Muhammad ﷺ and demonstrate how we can be confident that he indeed performed them.

This book will not resemble vague Western philosophical works filled with weak arguments debated in a vacuum. One of the most important realizations I have reached after years of study and debate is this: without a consistent standard of evidence and burden of proof, the entire discussion is meaningless.

Anyone can claim that a piece of evidence for miracles, evolution, or the moon landing is "compelling" or "definitive." Likewise, anyone can claim there is "no evidence" for one or more of these. But unless we establish a clear criterion for evidence—one that is applied equally to what skeptics already believe and to what they reject—there can be no objectivity in the discussion.

After the introduction, this book will dedicate specific chapters to each of the three topics in question. These chapters will examine the details, evidence, and logic behind miracles, evolution, and the moon landing, while exposing the double standards and inconsistencies in the positions of atheists, skeptics, and even Western academia.

It is important to note one thing. I am not claiming that the moon landing or all aspects of evolution are definitively false or provably false. In fact, even if you accept the moon landing or much of evolutionary theory, you should still have no difficulty engaging with this book—and, in fact, you will likely agree with much of what it presents.

This is because my purpose is not to disprove these beliefs outright, but to analyze and apply the same criteria used to defend them to the arguments for and against miracles.

When it comes to miracles and belief in general, I have noticed a problematic approach taken by many Muslims engaged in da'wah. They assume that every Islamic belief must be justified and proven, while skeptics and atheists often fail to justify most—if not all—of their own beliefs. I believe Muslims have absorbed this flawed approach from Christians, and it is an approach that obstructs a genuine search for truth.

These issues will be addressed in depth throughout the book. The reason this work—and my wider efforts on these subjects—are so crucial is simple: I am applying the very same standards demanded by non-Muslims and Western academia to their own beliefs. Without this equal application, anyone debating these topics is ultimately wasting their time.

This book represents an immense amount of hard work and dedication. Insha'Allah, I hope the facts and arguments presented here will open your eyes, just as they opened mine.

Introduction to Miracles — Definition, scope, possibility, burden of proof

Modern people often speak of miracles with a quiet smirk, as if the very word belongs to an age of ignorance long left behind. Yet the question of miracles is not a side issue—it sits at the heart of how we understand reality itself. If miracles are possible, the world is open to divine action and meaning; if they are impossible, then every claim of revelation, prophecy, or sacred history collapses with them.

This chapter examines what a miracle truly means, why belief in such events has come to be dismissed as irrational, and whether that dismissal is justified. We will ask whether miracles contradict science, whether evidence for the supernatural exists, and what the limits of human knowledge say about our ability to judge such events. By the end, it will become clear that the question of miracles is not about rejecting reason—but about expanding its boundaries to include all that reality might contain.

What Is a Miracle? — The Supernatural and Natural Law

When people speak of miracles, they usually have in mind an event that seems to go beyond the normal workings of nature—something that defies the ordinary patterns by which the universe functions. But a miracle need not always be an outright impossibility. It could be something that, while not violating any known physical principle, is highly unlikely under normal conditions that it stands apart from everything else we experience. In this chapter, we are not trying to fix a rigid definition or impose a single criterion for what counts as a miracle. The point is to clarify what might reasonably be meant by the term, and to consider how such events fit—or fail to fit—within our understanding of reality.

A common way of describing miracles, both by believers and skeptics, is that they are "violations of natural laws." The problem with this formulation is that it assumes something rather questionable: that such laws truly exist as absolute governing entities. The very concept of a "law of nature" has long been debated by philosophers. Some, like the scientific realists, hold that natural laws are objective features of the universe—stable principles that govern how things must behave under given conditions.[1] Others, such as empiricists and philosophers like David Hume or Nancy Cartwright, regard these "laws" not as binding rules but as generalizations drawn from experience, describing how things tend to behave, not why they must always do so.[2]

Now, questioning the absoluteness of natural laws does not mean that anything should be considered possible, nor that we should abandon rational constraints altogether. Rejecting the idea that every recurring pattern is an immutable "law" does not entail accepting absurdities such as things coming into existence from nothing, or violations of the Principle of Sufficient Reason. Rather, it means that what qualifies as a genuine natural law should itself be subject to evidence and philosophical scrutiny. If we find no empirical or rational basis for certain claims—such as something existing without cause or explanation—then such claims remain implausible. But likewise, if the realist view of natural law is correct, then the very existence of such laws implies an underlying order and purpose, and therefore the possibility of their deliberate suspension by the One who authored them. And if the empiricist view is correct, then a miracle would simply be an event that lies outside the patterns we have thus far observed—unusual, but not incoherent. In either case, miracles cannot be dismissed as meaningless or impossible.

Some philosophers have used the ambiguity of natural laws to argue against miracles. Hume defined a miracle as a violation of these laws and claimed that no testimony could ever outweigh humanity's consistent experience of their uniformity.[3] Some thinkers like Nancy Cartwright, however, argued that natural laws are not absolute truths but simplified descriptions of how the world usually behaves.[4] The debate remains unresolved—and rather than undermining miracles, it also weakens the skeptic's certainty. If laws are not metaphysically binding, then it makes little sense to claim that miracles are impossible.

The term supernatural is also often misunderstood. It is usually taken to mean something wholly separate from physical causation—God acting against nature. Yet in most theistic traditions, God sustains nature itself, so divine action is not a disruption but the ultimate cause of all continuity.[5] A miracle, then, could unfold through ordinary physical processes—only arranged in an extraordinarily precise or improbable way.

That idea may seem counterintuitive, but counterintuitive thinking is common in atheism as well. The emergence of consciousness from inert matter, morality from blind evolution, or unseen multiverses are all examples. Yet such claims are rarely treated as irrational, nor do atheists accept the same burden of proof for them. This shows that what counts as "rational" often reflects one's worldview more than the strength of evidence.

Thus, the question of miracles cannot be settled by invoking "laws of nature" or vague contrasts between the "natural" and the "supernatural." Both notions are far less clear-cut than modern discourse assumes. The reasonable position is to acknowledge that several coherent ways of understanding miracles exist—and none are ruled out a priori. The deeper issue, to which we now turn, is why belief in miracles—once accepted by philosophers, theologians, and scientists alike—has come to be seen as irrational in the modern world.

Why Miracles Are Considered Irrational Today — A Historical Shift

For most of human history, miracles were not considered irrational. The very idea of "the supernatural" being opposed to "the natural" would have seemed strange to classical thinkers. In ancient and medieval thought, the natural world itself was seen as purposeful and infused with divine intelligence. The universe was not an autonomous machine but an ordered system sustained by God or a higher principle. Within that worldview, extraordinary acts by the same divine will that governed ordinary events were perfectly coherent. A miracle was not a violation of reason but an expression of divine power beyond normal human expectation.[6]

In both Christian and Islamic traditions, this harmony between reason and revelation was well established. Thinkers such as Augustine, Aquinas, and Al-Ghazālī did not view miracles as breaking the order of nature but as rare demonstrations of God's will within it. What we call "natural laws" were understood as the consistent patterns of divine action—God's customs in creation. Miracles were exceptional acts of that same God, and thus no more irrational than the existence of the natural order itself. To deny the possibility of miracles would have seemed as unreasonable as denying the existence of nature's laws, for both arose from the same source.

This understanding began to change during the Enlightenment. The shift did not occur because of new evidence against miracles but because of a transformation in the very concept of reason and nature. The emerging philosophy of empiricism reduced knowledge to what could be observed and measured, and naturalism began to define reality as a closed system of physical causes. Within this new framework, divine action was no longer part of the natural order but something external to it. David Hume's famous essay on miracles solidified this attitude: by redefining a miracle as a "violation of natural law," he made disbelief in miracles a matter of definition rather than discovery.[7]

With the rise of Newtonian physics, this mechanistic outlook was reinforced. The universe came to be viewed as a vast clockwork governed entirely by impersonal forces. Once nature was conceived as a self-sufficient machine, divine action began to seem like an interference. The question was no longer whether miracles occurred, but whether they could occur at all in a closed, deterministic system. Thus, the modern rejection of miracles was not the conclusion of empirical investigation but the product of a new worldview—one that first excluded divine agency and then declared it irrational.

Miracles and the Existence of God

Miracles are often discussed in connection with God, and this association provides one natural way to understand them. However, it is not the only way to approach the topic. While the existence of God is frequently questioned or dismissed as irrational, this is a misconception. As I have argued in my book The Existence of God and the Irrationality of Atheism, belief in God is, in fact, the only rational explanation for the existence of the universe.[8] The very existence of a contingent universe, with its order and intelligibility, points to a necessary and purposeful cause.

If God is the Creator of the universe and its natural laws, then granting the existence of these laws implies that God also has the power to act beyond them. Miracles, then, are not violations of reason but manifestations of divine authority operating in harmony with the very source that sustains the universe. Just as it is rational to accept the universe itself as God's creation, it is equally rational to accept that God could intervene in extraordinary ways when He wills. Miracles, in this light, serve as signs that reveal the presence and power of the Creator rather than exceptions that contradict reason.

Even if we choose not to link miracles directly to God, the concept still retains its coherence. If natural laws are not absolute or do not exist independently, then there is nothing inherently implausible about extraordinary events occurring. The appearance of a miracle is only deemed "impossible" if one assumes that the universe is a rigid, law-bound mechanism entirely closed to any deviation. Without such an assumption, what we call miracles are simply extraordinary events—unusual, perhaps rare, but not logically incoherent.

Furthermore, the connection between miracles and God helps explain their significance. Miracles are not merely random anomalies; they are often purposeful events, intended to convey a message, demonstrate authority, or guide human understanding. Linking miracles to a conscious, purposeful Creator provides a rational framework for their occurrence, making them intelligible rather than arbitrary. At the same time, even a non-theistic perspective can acknowledge that miracles—events that depart from ordinary expectations—are conceptually possible wherever natural laws are not strictly fixed or absolute.

This understanding sets the stage for the next discussion: whether miracles are inherently irrational. By recognizing that miracles can be rationally coherent—either as acts of a divine Creator or as phenomena not bound by absolute natural laws—we can examine the deeper question of why they are often perceived as impossible or implausible in modern thought.

Is Belief in Miracles Inherently Irrational?

Having considered miracles in relation to God, we can now turn to the question of whether miracles are inherently irrational. This is not a question of evidence, but of intuition—the instinctive feelings or natural beliefs that humans have about the world. Many skeptics and atheists argue that miracles are intuitively or inherently implausible. However, this claim faces several serious challenges.

First, the belief in God and the notion that life is designed are themselves intuitive. Here is a reference regarding the belief in God being intuitive:

I tested both the Japanese and British children on the same tasks, showing them very accurate, detailed photographs of selected natural and man-made objects and then asking them questions about the causal origins of the various natural objects at both the scientific level (e.g. how did this particular dog become a dog?) and at the metaphysical level (e.g. how did the first ever dog come into being?). With the Japanese children, it was important to establish whether they even distinguished the two levels of explanation because, as a culture, Japan discourages speculation into the metaphysical, simply because it's something we can never know, so we shouldn't attempt it. But the Japanese children did speculate, quite willingly, and in the same way as British children. On forced choice questions, consisting of three possible explanations of primary origin, they would predominantly go for the word 'God', instead of either an agnostic response (e.g., 'nobody knows') or an incorrect response (e.g., 'by people'). This is absolutely extraordinary when you think that Japanese religion — Shinto — doesn't include creation as an aspect of God's activity at all. So where do these children get the idea that creation is in God's hands? It's an example of a natural inference that they form on the basis of their own experience. My Japanese research assistants kept telling me, 'We Japanese don't think about God as creator — it's just not part of Japanese philosophy.' So it was wonderful when these children said, 'Kamisama! God! God made it!' That was probably the most significant finding.[9]

Here is a reference of even Richard Dawkins admitting that life being designed is intuitive:

"Biology is the study of complicated things that give the appearance of having been designed for a purpose."[10]

Philosophical principles such as the Principle of Sufficient Reason (PSR) and other basic assumptions about order and causality are also intuitive. In all these cases, skeptics do not treat intuitions as decisive evidence for or against belief. Using intuition selectively to dismiss miracles is therefore inconsistent and intellectually dishonest.

Second, atheists and skeptics frequently hold beliefs that are counterintuitive, yet they do not apply the same burden of proof to these ideas. This inconsistency highlights the arbitrariness of claiming miracles are "intuitively" impossible.

A deeper problem emerges when we distinguish between two types of intuitions. Some intuitions are universal—they are shared across different cultures and historical periods. Examples include belief in God, moral sensibilities, or the recognition of purpose and design in the world. Other intuitions are subjective, varying from person to person or from culture to culture. The vast majority of humans throughout history have accepted miracles, or at least considered them plausible. This widespread acceptance indicates that the intuition against miracles is not universal, but rather subjective.

Subjective intuitions cannot serve as objective evidence for the impossibility of miracles. They differ among individuals and cultures, often contradicting each other. To claim that miracles are inherently irrational based solely on such intuitions is therefore unjustified. If anything, the historical and cross-cultural plausibility of miracles suggests that human reason and experience do not render them inherently impossible.

In short, the claim that miracles are counterintuitive relies on selective and culturally contingent intuitions. Once we recognize the distinction between universal and subjective intuitions, it becomes clear that miracles cannot be dismissed as inherently irrational. This understanding allows us to approach the study of miracles without presupposing their impossibility, focusing instead on their nature and significance.

Are Miracles Irrational Due to Lack of Evidence?

Another common objection to miracles is that they are supposedly irrational because there is insufficient evidence to support them. This objection is closely tied to the issue of the burden of proof. Skeptics often claim that, since evidence for miracles is lacking, the responsibility falls on those who argue in favor of miracles to prove their occurrence.

There are two significant problems with this reasoning. First, the burden of proof is determined before examining the evidence, not after. One cannot claim that the burden of proof lies with someone simply because evidence has not yet been presented. To do so is a logical misstep: it assumes the conclusion—that miracles are implausible—before evaluating the data. The same logic would, by this reasoning, apply inconsistently to other contested claims. For instance, skeptics of evolution or the moon landing do not claim the burden of proof shifts in the same way, despite ongoing debates and controversies.

Second, the claim of "no evidence" is inaccurate. There exists a considerable body of evidence for miracles, from historical accounts to documented instances in religious and historical literature. While the full comparison to the evidence for evolution or the moon landing will be addressed in later chapters, it is worth noting here that the case for miracles is far from negligible and, in many ways, presents a stronger historical and testimonial record than the evidence for either Evolution or the Moon landing.

In short, dismissing miracles on the grounds of a supposed lack of evidence is both logically flawed and empirically questionable. Before declaring miracles irrational, one must first fairly consider the evidence and the principles governing burden of proof, rather than assuming impossibility by default.

Is There Evidence of the Supernatural? — Prophecies as Miracles

A common tactic among atheists and skeptics is to deny that there is any credible evidence for miracles—or for the supernatural in general. They often attempt to undermine Christianity's claims to the miraculous and then assume that this dismissal extends to all religions. However, this approach cannot be applied to Islam. Islam stands on a distinct and far stronger foundation, supported by multiple, independent, and compelling proofs of its divine origin.

Among these are the Qur'an's unparalleled literary challenge, the numerous fulfilled prophecies of Prophet Muhammad ﷺ, and the presence of historically and scientifically accurate information in the Qur'an—details that were unknown in the pre-Islamic world. Together, these constitute a cumulative case for the truth of Islam that is unique among world religions.

Perhaps the simplest and most accessible of these evidences are the prophecies of the Prophet Muhammad ﷺ. These predictions—many of which were fulfilled precisely as foretold—demonstrate knowledge of the unseen that cannot be explained by chance, coincidence, or human insight. They serve as clear signs of divine revelation and as powerful evidence that Islam is from God.

Let us now examine two of these remarkable prophecies—prophecies that were not just fulfilled, but fulfilled in ways that no human in those times could have ever imagined.

The Prophet Muhammad ﷺ once made a prophecy that has now become one of the most visibly fulfilled prophecies in the modern world. He said:

He asked: "When will the Hour be?' He said: 'The one who is being asked about it does not know more than the one who is asking.' He asked: 'Then what are its signs?' he said: 'When the slave woman gives birth to her mistress' (Waki' said: This means when non-Arabs will give birth to Arabs") 'and when you see barefoot, naked, destitute shepherds competing in constructing tall buildings.' The Prophet (ﷺ) met me three days later and asked me: 'Do you know who that man was? I said" 'Allah and his Messenger know best.' He said: 'That was Jibril, who came to you to teach you your religion.'"[11]

Now, some try to downplay this prophecy by saying that people were already building tall structures throughout history—like the pyramids, or the Roman aqueducts, or the towers of Babylon. But that misses the entire point of the prophecy. It is not about tall buildings existing. It is about a very specific group of people—the barefoot, naked Bedouins—rising from absolute poverty to such wealth and power that they begin competing in building skyscrapers.

This is not just a prediction of architectural ambition—it is a prediction of socio-economic transformation on a massive scale.

The Bedouins were among the poorest people in the world. They had no cities, no infrastructure, and lived in tents while herding camels. The idea that such people would one day not only build modern cities, but compete in erecting the tallest buildings on Earth, was laughable at the time. It would be like predicting that the poorest shepherds in the

Himalayas would someday dominate global urban development. Yet this is exactly what happened.

Today, the tallest buildings on Earth rise from the deserts of Arabia. The Burj Khalifa stands as a gleaming monument to this very prophecy. Cities like Dubai, Doha, and Riyadh have transformed into futuristic skylines. And critically—it's not just about building—it's about competition. The Gulf States are in an ongoing race to outdo one another in the grandeur of their cities.

From Saudi Arabia's planned 170-kilometer-long city "The Line" to Dubai's endless skyline expansions, the competition is real, visible, and relentless.

Now ask yourself: how could anyone in 7th-century Arabia foresee such a future? There was no oil economy. There were no signs of coming wealth or civilization. These people were irrelevant to the major powers of the world.

There is no sociological trend, no historical precedent, and no economic indicator that could have led anyone to predict this. Except one: revelation from the Creator, who sees what no man can see.

Another extraordinary prophecy of the Prophet ﷺ is found in his statement:

It was narrated that Abu Hurairah RA said: "The Messenger of Allah ﷺ said: There will come a time when there will be no one left who does not consume Riba, and whoever does not consume it will nevertheless be affected by residue."[12]

This is not a vague religious warning—it is a precise description of a global financial reality that has become impossible to escape. At the time this was said, interest was banned by both Christianity and Islam. It was condemned by the religious authorities of the time as an immoral, predatory practice. Financial systems were simple, and those who practiced interest-based lending were a minority—typically wealthy moneylenders or ruling elites. The idea that everyone—rich and poor, religious and secular—would be consuming or affected by interest was inconceivable.

But today, the prophecy stands fulfilled before our eyes. Modern banking has institutionalized interest to such a degree that it forms the very backbone of the global economy. Credit cards, student loans, car financing, mortgages - everyday life is soaked in riba. Even governments are trapped in debt spirals. The entire financial system hinges on central banks adjusting interest rates to control inflation and spending. And for those who claim to avoid interest personally—it doesn't matter. Your currency, your purchasing power, your market prices—everything is affected by interest-based debt.

But one may ask: hasn't interest always existed? Yes, but not in the way the prophecy describes. The difference is total domination. Never before in history had riba become so universal and inescapable. One striking example is the invention of savings accounts, in which ordinary people earn interest simply by depositing money into a bank. This was unheard of in all ancient economies.

Historical records show that the first time people were paid interest for putting their money in banks was when the Bank of England was founded in 1694. While early versions of interest-based deposits appeared in Italian and Dutch banking in the 15th century, the system only became widespread centuries after the Prophet's death. This means that the concept of everyone—not just elites—consuming interest was something completely alien to his time.

The Prophet ﷺ did not merely predict that some people would engage in riba. He foretold a time when no one on Earth would be untouched by it—either as consumers or as indirect victims. This has never been the case in history—until now. Such precision, such foresight, from a man who never studied finance, never saw a bank, and never lived in a monetary economy like ours—this is nothing short of a miracle.

These and similar prophecies of the Prophet Muhammad ﷺ are massive proofs for the truth of Islam and reality of the supernatural. These are not vague predictions that can be twisted to fit any event, nor are they "safe bets" based on existing trends. They are often bold, precise, and at times seemingly impossible outcomes that were uttered in the middle of the desert by a man with no access to geopolitical influence, technological knowledge, or advanced education. He made claims that, had they turned out false, would have discredited him entirely. Yet history has vindicated his words time and again.

While such evidence may not fall under the category of physical miracles, they nonetheless constitute undeniable proofs of the supernatural. A fulfilled prophecy represents knowledge of future events that cannot be accessed by natural means. It involves information about contingent and specific occurrences that depend on countless variables, making accurate human prediction impossible. When such prophecies are numerous, detailed, and precisely fulfilled—as they are in the case of Prophet Muhammad ﷺ—the probability of coincidence or fabrication becomes vanishingly small.

These prophetic signs, therefore, serve as rational proof that a divine source of knowledge exists and that revelation is real. Once this is established, the possibility of miracles follows naturally. If the Qur'an is demonstrably from God, and if Prophet Muhammad ﷺ spoke with divine guidance, then there is no logical barrier to accepting other supernatural acts attributed to divine will. The recognition of true prophecy thus dismantles the skeptic's naturalistic assumptions and opens the door to acknowledging the broader reality of miracles.

In this way, the evidence for prophecy not only confirms the truth of Islam but also strengthens the philosophical foundation for belief in miracles themselves.

Psychic Phenomena and Parapsychology

The relationship between miracles and psychic phenomena is another area that deserves careful attention. Over the past century, there has accumulated an enormous body of evidence for various forms of psychic phenomena such as telepathy, clairvoyance, psychokinesis, and precognition.

- Telepathy refers to the transmission of thoughts or information directly from one mind to another without the use of known sensory channels.

- Clairvoyance is the ability to perceive objects, people, or events beyond the range of ordinary perception.

- Psychokinesis (or telekinesis) is the direct influence of mind upon matter without physical contact.

- Precognition is the perception or knowledge of future events before they occur.

The evidence for these abilities is not limited to anecdotes or folklore but is drawn from thousands of carefully documented experiments, case studies, and cross-cultural reports. Compelling evidence for psychic phenomena has existed for a long time, and even many skeptics who do not personally accept them as genuine have conceded that the data supporting their existence is extensive and cannot be dismissed lightly. Here are some references on this topic from skeptics.

The following reference is from the skeptic Carl Sagan who is talking about interesting developments relating to psychic phenomena which deserve further research:

"At the time of writing there are three claims in the ESP field which, in my opinion, deserve serious study: (1) that by thought alone humans can (barely) affect random number generators in computers; (2) that people under mild sensory deprivation can receive thoughts or images "projected" at them; and (3) that young children sometimes report the details of a previous life, which upon checking turn out to be accurate and which they could not have known about in any other way than reincarnation."[13]

In 1986, after much debate, a skeptic Ray Hyman and a parapsychologist Charles Honorton jointly published an article on psychic phenomena where the skeptic agreed that the research cannot be explained as a result of underreporting by the parapsychologists even though the skeptic did not fully agree to these results being proof of psychic phenomena:

We agree that there is an overall significant effect in this database that cannot reasonably be explained by selective reporting or multiple analysis. We continue to differ over the degree to which the effect constitutes evidence for psi, but we agree that the final verdict awaits the outcome of future experiments conducted by a broader range of investigators and according to more stringent standards.[14]

After some years and advancement in the field, the skeptic Ray Hyman later remarked on further evidence being discovered for psychic phenomena:

"The case for psychic functioning seems better than it has ever been. . . . I also have to admit that I do not have a ready explanation for these observed effects."[15]

This was the situation in 1996 but the thing to understand is that the evidence for psychic phenomena has been present for a very long time. Back in 1955, Dr. George Price, then a research associate at the Department of Medicine at the University of Minnesota, published an article in the prestigious journal Science that began:

Believers in psychic phenomena . . . appear to have won a decisive victory and virtually silenced opposition. . . . This victory is the result of careful experimentation and intelligent argumentation. Dozens of experimenters have obtained positive results in ESP experiments, and the mathematical procedures have been approved by leading statisticians. . . . Against all this evidence, almost the only defense remaining to the skeptical scientist is ignorance.[16]

It is not known about because of the deception and shifting of goalposts propagated by Western Academics. The skeptic Marcello Truzzi summed up the history of laboratory parapsychology:

As proponents of anomalies produce stronger evidence, critics have sometimes moved the goal posts further away. . . . To convince scientists of what had merely been supported by widespread but weak anecdotal evidence, parapsychologists moved psychical research into the laboratory. When experimental results were presented, designs were criticized. When protocols were improved, a "fraud proof" or "critical experiment" was demanded. When those were put forward, replications were demanded. When those were produced, critics argued that new forms of error might be the cause (such as the "file drawer" error that could result from unpublished negative studies). When meta-analyses were presented to counter that issue, these were discounted as controversial, and ESP was reduced to being some present but unspecified "error some place" in the form of what Ray Hyman called the "dirty test tube argument" (claiming dirt was in the tube making the seeming psi result a mere artifact). And in one instance, when the scoffer found no counter-explanations, he described the result as a "mere anomaly" not to be taken seriously so just belonging on a puzzle page. The goal posts have now been moved into a zone where some critics hold unfalsifiable positions.[17]

There is an interesting parallel between psychic phenomena and miracles. Both involve occurrences that seem to go beyond the known limits of natural law, and both are often dismissed by atheists and skeptics as belonging to the same category of superstition. However, parapsychologists—the scientific researchers who study psychic phenomena—generally argue that psychic phenomena should not be classified as "miracles."

Parapsychologists maintain that psychic phenomena occur far more frequently and consistently than miracles are reported to. Because such events are found in diverse cultures and time periods, they argue that these phenomena are part of the natural order rather than singular, divinely orchestrated interventions. Moreover, psychic effects can be tested and measured under experimental conditions, with results that have been replicated across studies. This scientific testability, they suggest, differentiates psychic phenomena from unique miraculous events that cannot be reproduced on demand.

Another point parapsychologists emphasize is that psychic phenomena likely operate according to natural principles not yet understood. Just as forces such as magnetism or electricity were once regarded as mysterious but are now well-integrated into scientific understanding, psi effects may one day be explained through extensions of physics, biology, or consciousness studies. For this reason, they view psychic phenomena as natural rather than supernatural occurrences. Finally, many parapsychologists interpret these phenomena as latent or exceptional human capacities—extensions of consciousness and perception—rather than as acts of divine agency.

From a traditional theological standpoint, however, psychic phenomena could still constitute miracles in the broader sense—acts that transcend our ordinary causal framework and point to realities beyond material explanation. If psychic phenomena truly occur—and the evidence strongly indicates that they do—then the plausibility of miracles increases dramatically. Psychic phenomena show that the natural world is not closed and that the mind can, in some way, influence matter and access information beyond physical boundaries. This demonstrates that immaterial causation is possible, directly challenging the materialist assumption that all effects must have purely physical causes.

Furthermore, the reality of psychic phenomena bridges the conceptual gap between divine and human agency. If the human mind can, in limited and measurable ways, exert influence beyond physical mechanisms, then divine influence in the world becomes far more intelligible. The principle that consciousness can affect the physical world opens the door to the possibility that a higher consciousness can likewise act within it.

In short, if psychic phenomena are real—and the cumulative evidence strongly suggests they are—then the notion of miracles is not only credible but expected. Both point toward a universe that is richer, more mysterious, and more open to transcendent realities than materialism allows.

Do We Need to Know How Miracles Occur?

A common objection raised against miracles is that we do not know how they occur. Skeptics argue that since we cannot explain the mechanism by which a miracle happens, we should not accept its occurrence. This reasoning, however, is deeply flawed. It confuses understanding the cause of an event with recognizing that the event took place.

To deny the reality of something merely because we do not know how it operates is intellectually inconsistent. Human knowledge has always expanded by first acknowledging facts and only later discovering the mechanisms behind them. People observed gravity, magnetism, and disease long before understanding the laws that govern them. The world was full of functioning realities—like lightning, infection, and heredity—whose workings were unknown for centuries. Ignorance of how something occurs does not mean it does not occur.

This objection also falls into what is often mischaracterized as the "God of the gaps" fallacy. The real "gap" here lies not in the believer's reasoning, but in the skeptic's logic. The believer does not invoke God because there is no explanation, but because the nature of the event itself—its context, timing, purpose, and defiance of natural causation—points rationally toward divine intervention. To dismiss all unexplained phenomena as placeholders for ignorance is to commit the opposite fallacy: the materialism of the gaps—the blind faith that science will one day explain everything naturally, regardless of the evidence.

Furthermore, this standard is applied selectively. Scientists frequently accept phenomena as real even when the precise mechanisms are unknown. For example, in evolutionary biology—often cited by skeptics as the paradigm of scientific certainty—there remain vast gaps in our understanding of how major transitions occurred. We know that species exist and have changed over time, but the detailed causal steps for many key evolutionary events remain speculative.

Consider a few examples:

- The origin of life (abiogenesis): There is still no demonstrated natural pathway from non-living chemicals to a self-replicating cell. Dozens of theories exist, yet none have bridged the gap experimentally.

- The Cambrian explosion: The sudden appearance of nearly all major animal body plans in the fossil record lacks a clear evolutionary explanation. The incremental transitions predicted by Darwinian theory are still missing.

- Consciousness: Even leading neuroscientists admit that no physical theory adequately explains how subjective awareness arises from matter—a mystery far greater than any miracle claim.

- The origin of complex biological systems: The precise evolutionary pathways for structures like the bacterial flagellum, the eye, or the blood-clotting cascade remain highly contested, with only speculative reconstructions offered.

In each of these cases, Evolutionists accept the data as real despite not fully knowing how the processes occurred. They do not reject evolution outright because of incomplete knowledge—they simply acknowledge the limits of current understanding. Yet when it comes to miracles, this attitude vanishes. The skeptic suddenly demands complete mechanistic explanation before admitting even the possibility of occurrence.

This double standard reveals the bias, not the believer's inconsistency. If lack of mechanism does not invalidate the facts of evolution, it cannot logically invalidate the occurrence of miracles either. One may not know how God heals the sick, splits the sea, or fulfills a prophecy—but to reject the event solely on that basis is to confuse ignorance of process with denial of evidence.

In reality, the question "how do miracles happen?" is misplaced. By definition, a miracle involves divine agency operating beyond the limits of known natural laws. Understanding its mechanism is not necessary to affirm its occurrence—just as understanding the ultimate mechanism of consciousness or gravity is not necessary to know that they exist. The appropriate question is not how a miracle happens, but whether there is credible evidence that it did.

Believing in the Possibility of Miracles

A frequent tactic employed by atheists and skeptics is to demand that believers first prove that miracles are possible before even considering any evidence for their occurrence. This demand, however, is both unreasonable and inconsistently applied. If we were to follow this same standard, we would have to first prove that evolution or the moon landing were even possible before examining the evidence for them. Yet skeptics never hold those positions to the same scrutiny—and for good reason: the "possibility" of an event cannot be proven in the abstract without reference to its evidence.

In fact, the very concept of "proving possibility" is philosophically incoherent. How can one demonstrate that something is possible without first demonstrating that it either did or could occur under certain conditions? To ask for proof of possibility before examining evidence is to demand a conclusion before the argument begins. The only meaningful way to assess possibility is by looking at the evidence itself. If credible evidence exists that a miracle occurred, then its possibility is automatically established.

This point becomes clearer when we apply it consistently. If a skeptic were to demand that evolutionists first prove that macroevolution is possible before examining fossils or genetics, or that historians first prove that space travel is possible before studying the moon landing, the demand would be dismissed as absurd. Yet when it comes to miracles, this very same fallacy is routinely invoked as if it were sound reasoning.

The truth is that the possibility of almost anything cannot be proven apart from the thing itself. We did not know heavier-than-air flight was possible until it was done. We did not know splitting the atom was possible until it happened. We do not prove possibility in the abstract—we recognize it through demonstration. To ask for proof of possibility prior to examining evidence is not critical thinking; it is a means of avoiding it.

Moreover, if skeptics wish to be consistent, they must apply their standard equally to their own beliefs. Evolutionary theory, for instance, includes countless phenomena that even leading scientists admit they cannot fully explain. There are vast gaps in understanding how life originated, how consciousness arose, or how complex biological systems developed from simpler forms. These are not minor details—they are foundational mechanisms. If we applied the skeptic's "prove possibility first" standard to these issues, we would have to suspend belief in evolution entirely until every mechanism was known. Yet no rational person would make such a demand, because science progresses by evidence, not by hypothetical possibility.

The inconsistency deepens when we consider atheistic metaphysics. Many atheists argue that the universe can emerge from nothing or that an actual infinity of past events is possible. They insist that these are "plausible" or "reasonable" positions because they involve no logical contradictions. Here is a reference regarding this claim from Atheists:

So one of the first things that mathematicians come to appreciate is that infinite collections do have different properties from finite collections. Things that we absolutely take for granted in the finite case simply don't carry over to the infinite case. And if it's the finite case that is providing us with our basic intuitions then that means that some of the results about infinite sets are going to be counterintuitive. But that's just what you would expect. Why should the finite and the infinite be the same? As long as we're prepared for these differences, again the very worst that it will be is counterintuitive. It won't be contradictory. It won't be inconsistent.[18]

Yet this same reasoning, if applied fairly, makes miracles equally possible. There is no logical contradiction in the idea that a transcendent Creator could intervene in creation. A miracle may be unusual or beyond ordinary experience, but it is not illogical.

If one can believe that matter arose from non-matter, that consciousness emerged from lifeless chemicals, or that infinite regress is metaphysically coherent, then believing that God can heal the sick, foretell the future, or part the sea is far less of a stretch. In fact, miracles are far more intellectually consistent, for they posit a rational cause—divine will—behind extraordinary effects.

Thus, the demand to "prove miracles possible" before considering evidence is a smokescreen. It is not about logic but about bias. Once we apply consistent standards across all domains—science, history, and philosophy—the possibility of miracles stands on equal, if not stronger, rational ground. The real question is not whether miracles are possible, but whether there is credible evidence that they have occurred. And as we will see, such evidence is abundant.

Burden of Proof and Standards of Rationality

A recurring theme in skeptical argumentation is the claim that believers in miracles bear the entire burden of proof. According to this view, the skeptic need not defend his own position; he merely waits for the believer to fail. Yet this standard is never applied consistently. When it comes to evolution or the moon landing, the same skeptics insist that those who deny these events carry the burden of proof. In other words, the burden shifts conveniently depending on which belief they want to defend.

If the justification for this inconsistency is intuition—that evolution and the moon landing "feel" more plausible—then such reasoning is shallow and self-defeating. Both subjects contain elements that defy intuition. The idea that blind, purposeless mutations produced consciousness and moral reasoning is profoundly counterintuitive. Likewise, the notion that humans traveled hundreds of thousands of kilometers through space, landed on a distant celestial body, and returned safely with 1960s technology, defies ordinary human experience far more than most people realize. Intuition, then, is a poor measure of truth, and certainly not a rational basis for assigning burden of proof.

If the justification is that "miracles are not happening right now," then that reasoning collapses just as quickly. Evolution is not happening before our eyes either—not on the scale or complexity claimed by Darwinists. Nor is the moon landing repeating itself today. Both are historical claims about past events, just as miracles are. If temporal distance removes miracles from consideration, it must also remove evolution and the moon landing.

Others argue that the difference lies in evidence. But this too misunderstands how the burden of proof works. The burden of proof is assigned before examining evidence, not after. It is a matter of epistemic fairness, not convenience. To claim that believers in miracles bear the burden of proof because the evidence is lacking is circular reasoning—it assumes the very conclusion under dispute. Moreover, the supposed "evidence" for evolution and the moon landing is far from beyond criticism. Both rely heavily on interpretive frameworks, unverified assumptions, and institutional trust, rather than direct observation or reproducible data.

There is a deeper problem still. Skeptics often smuggle into the discussion a hidden rule: that if any alternative explanation for miracle evidence exists—no matter how speculative—then it ceases to count as evidence for a miracle. This is an impossible standard, one that would invalidate every field of human inquiry if applied consistently.

Take the moon landing, for instance. There are alternative explanations for every piece of evidence—photographs, videos, rock samples, and telemetry data—all of which have been challenged by those who doubt the official narrative. Likewise, evolution has alternative interpretations for its fossil record, genetic similarities, and vestigial structures. In fact, even mainstream scientists acknowledge unresolved debates and competing explanations within evolutionary theory itself. Yet skeptics do not dismiss these fields as irrational simply because alternative hypotheses exist.

Why, then, should miracles be judged differently? Evidence does not cease to exist merely because an alternative story can be imagined. In every discipline, the standard is not "no other explanation possible," but "which explanation best accounts for the data." By that same standard, many miracle claims—especially those supported by multiple witnesses, historical continuity, or fulfilled prophecy—are far more compelling than the speculative reconstructions offered in evolutionary biology or certain aspects of space history.

The double standard is plain: when evidence supports their worldview, skeptics allow interpretation, trust, and inference; when evidence challenges it, they demand impossibility-level proof. This is not rational inquiry—it is selective skepticism.

In truth, a fair standard of rationality requires consistency. If the evidence for evolution and the moon landing can be accepted despite gaps, interpretation, and reliance on authority, then evidence for miracles—supported by testimony, documentation, and coherence with theism—deserves at least equal consideration. The burden of proof, therefore, does not rest solely with the believer. It rests equally with anyone making truth claims about reality.

When measured by a consistent standard, miracles are not the irrational outlier skeptics make them out to be. They are part of the same human quest for truth—one that demands honesty, humility, and the courage to follow evidence wherever it leads.

Conclusion

Throughout this chapter, we have seen that the skepticism directed toward miracles rests not on reason, but on selective and inconsistent standards. The objections raised—whether about lack of evidence, supposed impossibility, or misplaced demands for burden of proof—collapse under scrutiny. When the same logic is applied consistently to other widely accepted beliefs such as evolution or the moon landing, the skeptic's criteria prove arbitrary and self-serving.

The claim that miracles are irrational because we do not know how they occur ignores the fact that countless accepted phenomena are still unexplained in their mechanisms. The insistence that we must first prove miracles possible before considering evidence betrays a misunderstanding of how rational inquiry works. And the attempt to shift the entire burden of proof onto believers in miracles while excusing secular claims from the same demand reveals a deep philosophical bias rather than intellectual rigor.

At its core, the rejection of miracles is not rooted in logic but in presupposition. It assumes, without justification, that the natural world is a closed system and that divine intervention is impossible by definition. But once this circular assumption is dropped, the case for miracles becomes not only reasonable but compelling. There is abundant evidence—from fulfilled prophecy to verified events and credible testimony—that points to realities beyond material explanation.

The next chapter will explore the general logic behind arguments for miracles—how such arguments are structured, what makes them rational, and how they fit within a coherent understanding of reality. Rather than merely responding to skepticism, I will outline a positive, systematic framework that shows why belief in miracles is not only possible, but intellectually necessary once the evidence and logic are fairly examined.

Miracles: The Evidential Foundations — testimonies, verification, and philosophical foundations

Miracles occupy a unique place in human thought — standing at the crossroads of faith, history, and philosophy. For skeptics, they represent a challenge to a purely naturalistic worldview; for believers, they serve as powerful confirmation of divine truth. Yet before any discussion of specific miracles can be meaningful, one must first examine the foundations upon which belief in miracles rests: the credibility of testimony, the standards of verification, and the philosophical assumptions underlying how we judge extraordinary claims.

This chapter explores these foundational questions in depth. It begins by noting that reports of miracles are not confined to one religion or culture but are universal across human history — a fact that demands explanation. From there, it considers the role of testimony as a form of evidence and asks how trustworthy eyewitness accounts can be, especially in historical contexts. We will also explore how historians, scientists, and legal systems treat corroboration and reliability, and whether those same principles can be applied to reports of miracles.

Next, the chapter turns to an often-overlooked issue: the widespread acceptance of historical events and phenomena that are no longer observed today. This invites a crucial question — why do we accept many historical claims on testimony alone, yet treat miracle reports differently? Finally, the discussion culminates with an examination of the Islamic historical tradition, showing that its methods of transmission, verification, and authentication are among the most rigorous in the world — providing a uniquely strong evidential foundation for belief in miracles.

In doing so, this chapter seeks not merely to defend the possibility of miracles, but to demonstrate that belief in them can rest on firm rational and evidential grounds.

A Consistent Testimony Across Civilizations

From the dawn of recorded history to the present day, humanity has been surrounded by accounts of miraculous events — moments where the ordinary order of nature seems to disappear. Every civilization, no matter how isolated or diverse in its worldview, has reported such occurrences. Ancient Egyptian inscriptions speak of divine signs; the Vedas and Buddhist chronicles recount supernatural interventions; Greek and Roman historians recorded healings and prophetic visions; Chinese annals speak of celestial wonders; and the oral traditions of Africa and the Americas preserve stories of miracles long before contact with other cultures. Even in our own supposedly "enlightened" age, reports of near-death experiences, unexplainable recoveries, and prophetic dreams persist across continents.

This universality is far from trivial. It reveals that the recognition of the miraculous is not a cultural accident or a by-product of superstition, but a constant of human experience. When a phenomenon appears independently across time, geography, language, and belief system, the burden of explanation shifts. It is no longer enough to dismiss miracle reports as the product of ignorance or wishful thinking — for such dismissal would imply that every civilization and generation has, on precisely this issue, been entirely deluded in the same way. That level of collective human error strains all plausibility.

If all human societies have, in one form or another, witnessed and attested to events that seem to transcend nature, the reasonable question is not "why do so many people believe in miracles?" but rather "why should we assume they are all wrong?" When independent witnesses across history and culture consistently report similar types of experiences — healings, answered prayers, prophetic insights, deliverance from danger, and divine signs — it suggests that something real lies behind the pattern.

Skeptics sometimes argue that human beings are simply prone to misinterpret extraordinary but natural occurrences as supernatural. Yet even this objection, if true in a few cases, fails to explain the persistence, depth, and universality of miracle claims. If human perception were that unreliable, we would expect universal confusion in every area of life — not just when people describe experiences that point to the divine.

The universality of miracle reports, therefore, stands as a formidable piece of evidence in its own right. It points to a shared human encounter with realities that transcend the material, experienced and recognized by countless individuals and communities across the ages. To deny this testimony outright is not an act of rational skepticism, but of selective disbelief — an unwillingness to accept what humanity, almost unanimously, has always known.

Of course, the universality of miracle reports does not mean that every individual claim is true or beyond question. Human error, exaggeration, and misinterpretation can and do occur. Yet this observation proves little, for the same is true in every field of human inquiry. Countless scientific theories have been proposed and later abandoned; numerous historical accounts have been corrected or refined. But no one concludes from this that science itself is false or that history cannot be trusted. The proper response is not to dismiss the entire category, but to sift the evidence with care and discernment. Likewise, while some miracle reports may be mistaken, the sheer volume, consistency, and persistence of such testimonies throughout human history indicate that many cannot be so easily explained away.

To reject them all wholesale would be as irrational as rejecting all science because some scientists have erred.

The importance of testimony as evidence

Virtually everything we know about the world beyond our immediate senses comes to us through the testimony of others. We rely on the words and records of countless people—scientists, historians, journalists, teachers, and ordinary witnesses—for our understanding of reality. No one has personally verified the existence of every country, every past event, or even most scientific facts we routinely accept as true. Our trust in such knowledge rests, in large part, upon the credibility and consistency of testimony.

This reliance is not a weakness but a fundamental feature of human knowledge. Even in the most rigorous scientific contexts, testimony is essential. A physicist accepts the results of experiments published by colleagues; a historian trusts ancient chroniclers; a judge depends on witness statements. Modern society would collapse if we demanded personal verification for every fact we believe. The notion that testimony is somehow inferior to direct observation is not only impractical—it is philosophically untenable.

For centuries, however, some philosophers—most famously David Hume—regarded testimony with suspicion, treating it as a lesser source of knowledge that must be constantly checked against reason and empirical evidence.[19] This skepticism was shaped by the Enlightenment's emphasis on individual perception and experiment. Yet not all thinkers agreed. Thomas Reid, one of Hume's contemporaries, argued that trust in human testimony is a natural and rational principle of the human mind—so basic that rejecting it would make knowledge itself impossible.[20]

In recent decades, this debate has taken on new life within analytic philosophy. Contemporary scholars such as C. A. J. Coady, Elizabeth Fricker, and Jennifer Lackey have demonstrated that testimony is not merely a derivative or secondary form of knowledge but a primary and indispensable one.[21] In reality, most of what we believe—even in science—depends on the word of others whom we have no reason to distrust. The denial of testimony as a valid epistemic source would therefore render the vast majority of human knowledge unintelligible.

Once this is recognized, it becomes clear that testimony cannot be dismissed merely because it reports extraordinary events. The question is not whether a claim is transmitted through testimony, but whether the testimony itself is credible, consistent, and well-corroborated. If multiple trustworthy witnesses independently report the same event, if their accounts are coherent and align with known facts, then their collective word carries genuine evidential force.

Miracles, by their very nature, are historical and experiential—they occur to people, not under laboratory conditions. The primary evidence for them, therefore, will always be testimony. To reject miracle reports on the sole basis that they are testimonies is to commit a category error. If testimony is a valid form of evidence in law, science, and history, it cannot be arbitrarily excluded when it points toward the divine. The proper standard is not prejudice against testimony, but scrutiny of its quality.

Recognizing this restores balance to the discussion. The testimony of sincere, intelligent, and independent witnesses—especially when recorded, transmitted, and verified through reliable means—has the right to be considered as evidence. Indeed, without testimony, almost all human knowledge would vanish. To accept testimony when it supports our preferred conclusions but reject it when it points to miracles is not rational skepticism; it is selective disbelief masquerading as reason.

The reliability of eyewitness testimony

Eyewitness testimony is among the most pervasive forms of human evidence. It forms the backbone of social life, law, journalism, and historical record. Every day, countless transactions, investigations, and judgments depend upon people reporting what they saw and heard. If such testimony were generally unreliable, human cooperation and knowledge itself would collapse. The simple fact that our societies function—contracts are honored, histories are written, news is trusted, and justice is administered—demonstrates that eyewitness accounts are ordinarily dependable. This vast and continuous confirmation constitutes overwhelming empirical evidence for the practical reliability of testimony in ordinary life. When people describe events that are subsequently verified, the reasonable conclusion is not that such accounts are exceptional, but that testimony, in general, works.

Yet within modern psychology and law, debate continues regarding the limits of eyewitness accuracy. Since the 1970s, scholars such as Elizabeth Loftus have demonstrated that memory is malleable and vulnerable to suggestion. Experiments such as the famous "Reconstruction of Automobile Destruction" study showed that subtle variations in question wording could influence how participants recalled an event.[22] This and similar work revealed that leading questions, exposure to misinformation, and social pressure can alter details in a witness's memory. These findings led to justified reforms in legal procedure, emphasizing caution when evaluating identification evidence or testimony gathered under suggestive conditions.

However, popular presentations of these findings have often gone far beyond what the empirical literature warrants. While it is true that human memory can be distorted, it is not true that it is generally unreliable. As psychologists Gary Wells and Elizabeth Olson observed in a comprehensive review, eyewitness evidence is fallible but highly informative when obtained under proper conditions.[23] Similarly, the U.S. National Research Council's major report, Identifying the Culprit, does not call for abandoning eyewitness evidence but rather for improving methods of collection and evaluation to safeguard its accuracy.[24]

More recent scholarship has gone further in restoring balance to the discussion. Researchers such as John Wixted and colleagues have emphasized that when eyewitness memory is uncontaminated by suggestive questioning or procedural errors, it performs remarkably well. Their review, "Rethinking the Reliability of Eyewitness Memory," published in Perspectives on Psychological Science, argued that eyewitness memory is, in fact, one of the most reliable forms of evidence when evaluated under fair and controlled conditions.[25] The crucial point is that most documented inaccuracies arise not from inherent unreliability but from identifiable external factors that can be prevented.

In short, the modern academic consensus is not that eyewitness testimony is worthless, but that it must be handled carefully. Its reliability is conditional, not negligible. Empirical studies consistently show that when conditions are fair—when witnesses are not pressured, misled, or exposed to misinformation—their recollections are typically accurate. Field studies, such as Yuille and Cutshall's research on real-life eyewitnesses to a shooting, found that witnesses retained accurate memory of central details even months after the event.[26] These findings confirm what ordinary experience already tells us: testimony is not infallible, but it is reliable enough to form the foundation of knowledge in both everyday and institutional contexts.

When this understanding is applied to historical and religious testimony, including reports of miracles, the situation changes in significant ways. The standard objections to eyewitness evidence—drawn from the legal or forensic setting—do not map neatly onto the contexts in which ancient miracle reports emerged. In modern trials, eyewitnesses are often subject to leading questions, adversarial incentives, repeated interviews, and contamination from media or other witnesses. Such conditions are now known to introduce systematic distortions in memory. But these are not the conditions under which the earliest miracle reports were typically recorded. Ancient witnesses of extraordinary events were not participants in adversarial courtroom procedures designed to secure convictions; rather, their testimony was often public, communal, and uncoerced.

Moreover, much of the empirical criticism of eyewitness evidence concerns identification tasks—choosing a suspect's face from a lineup—where visual similarity and suggestive context cause confusion. Miracle reports, by contrast, usually describe events, not persons: healings, visions, celestial signs, or sudden changes in the natural order. The psychological errors involved in misidentifying a face from a small group are not the same as those involved in misremembering a dramatic and singular occurrence. Nor were early witnesses bombarded by modern media or repeatedly interviewed under subtly suggestive questioning, both major causes of distortion in laboratory experiments.

In oral and religious cultures, further safeguards existed that modern critics often overlook. Important events were preserved through communal memory, ritual retelling, and repetition. Far from distorting accounts, these mechanisms tend to stabilize core narratives. In some cases, as with early Islamic transmission practices or the memorization techniques of ancient historians, such collective preservation operated with a precision and consistency that modern empirical work would recognize as forms of memory reinforcement. The communal nature of transmission means that errors could be corrected through shared recitation and mutual oversight—very different from the isolated witness in a police interview room.

There is a very crucial point to keep in mind though. Disagreements among eyewitnesses about peripheral details do not undermine the core reality of an event. In virtually every domain of life, multiple observers describe the same incident with minor variations—one recalls the color of a car differently, another the precise time, a third forgets a small sequence. Yet all agree that the central event occurred. Courts, historians, and journalists all recognize this as normal: human perception captures events selectively, not with photographic precision. Such variation is actually a mark of authenticity, since perfectly uniform testimony would suggest collusion rather than independent observation. The crucial epistemic point is that divergence in incidental details does not erase the shared evidence of occurrence. Even where witnesses differ about how a miracle happened, the convergence that something extraordinary took place remains powerful and evidential.

The practical conclusion, therefore, is not that all miracle reports should be accepted uncritically, but that they cannot be dismissed wholesale by appealing to modern debates about eyewitness unreliability. Those debates concern a narrow, highly specific context that differs profoundly from the communal, historical, and religious settings in which reports of the miraculous arose. The fair epistemic posture is to apply the same criteria historians use for all testimony: proximity to the event, internal coherence and reliability of the witnesses and sources. When such criteria are satisfied—as they often are in major historical miracle traditions—eyewitness testimony deserves serious evidential consideration.

The broader philosophical point is clear. If testimony can yield reliable knowledge about crime scenes, accidents, and scientific phenomena, there is no principled reason to exclude it from the domain of miracles. The universality and practical success of testimony in human life is itself strong evidence that it functions as a trustworthy channel of truth. What modern research ultimately shows is not that we should distrust human witnesses, but that we should understand and interpret them properly.

The reliability of historical records

If eyewitness testimony forms the foundation of knowledge in everyday life, historical records are its enduring monument. Written history is, in essence, the preservation and transmission of testimony across time. The letters, chronicles, and documents that inform our understanding of the past are simply testimony recorded in more durable form. Thus, any argument that undermines the reliability of testimony would, by extension, collapse our confidence in the entire edifice of historical knowledge. Yet no serious historian doubts that we can know the past. The success of the historical enterprise itself is evidence that recorded human testimony, when critically evaluated, is a reliable source of truth.

Historians, of course, do not accept every document at face value. Over centuries, they have developed sophisticated methods for assessing sources: authorship, proximity to events, internal consistency, corroboration, and alignment with known historical or archaeological facts. These criteria are what enable historians to separate probable truth from legend or propaganda. The fact that such techniques consistently yield coherent, verifiable accounts of the past is strong evidence that historical documents can indeed transmit reliable information. The accuracy of modern reconstructions of ancient civilizations, battles, or dynasties—often confirmed by archaeology or independent records—demonstrates the success of this approach.

Critics sometimes point out that historical writers can be biased or selective. That is true, but selectivity does not equate to falsehood. Every source reflects a perspective; what matters is whether the core facts withstand cross-examination. Modern historiography has long acknowledged that bias can coexist with factual reliability. For example, the Roman historians Tacitus and Josephus each wrote with clear political and cultural biases, yet their works remain invaluable and historically dependable because they accurately describe verifiable events and persons. The task of the historian is not to reject biased accounts, but to interpret them critically. The same standard should apply to historical miracle reports: bias or religious motivation may shape interpretation, but they do not automatically invalidate the occurrence of the event itself.

Another persistent error is the assumption that temporal distance automatically diminishes reliability. Skeptics often treat ancient records as if age alone discredits them, while simultaneously trusting ancient sources on secular matters. Yet almost everything we know about figures like Alexander the Great, Socrates, or Julius Caesar comes from works written decades or centuries after their lifetimes. We have far fewer primary documents for these figures than for many recorded miracles, yet historians do not doubt their existence or their principal deeds. The reason is simple: historical knowledge is probabilistic, not absolute. It rests on the cumulative weight of consistent testimony, not on laboratory replication. To demand "scientific proof" for historical claims, including miracles, is to confuse the methods of science with those of history.

Furthermore, it is essential to distinguish between mythic or symbolic storytelling and historical record-keeping. Many miracle reports do not arise from mythic genres but from historical, legal, or biographical writings with clear temporal and geographical references. Early Muslim historians, for instance, treated the recording of events—including miracles and extraordinary occurrences—with remarkable methodological rigor, tracing chains of transmission and assessing the credibility of narrators.

When properly evaluated, historical records often prove far more trustworthy than modern skepticism allows. The same principles that allow historians to reconstruct battles, migrations, and inventions should also guide the assessment of miracle reports. If testimony is the lifeblood of everyday knowledge, then written testimony—the historical record—is the memory of humanity itself. To dismiss that memory selectively, accepting it for ordinary events while rejecting it for extraordinary ones, is not a mark of critical thinking but of inconsistency. The reliability of historical record stands or falls together: if we can know anything about the past, we can also know whether certain extraordinary events were seen, remembered, and recorded.

Belief in other things which do not happen now

A common objection raised by skeptics is that miracles cannot be trusted because we do not see them occurring today. The assumption is that if something truly happened, it should still be happening now. Yet this reasoning is unsound. Human knowledge of the past often depends on testimony and written record rather than present observation. There are many things we confidently accept as true, even though they no longer happen—or at least have not happened for centuries—and our knowledge of them comes primarily through the reports of earlier generations.

Consider the accounts of ancient wars and political upheavals that shaped entire civilizations. No one today doubts that the Persian, Roman, or Mongol empires waged massive campaigns that reshaped history. Yet our certainty of these events comes from chronicles, eyewitness accounts, and the writings of historians—not from direct observation. Similarly, no one now witnesses

gladiatorial games, public sacrifices, trials by ordeal, or duels to the death, though these were once common and vividly described by contemporaries. These events are accepted because the historical record of them is abundant, coherent, and independently corroborated through testimony.

Even within the modern period, there are countless examples of once-common occurrences that no longer happen but are not doubted: for instance, medieval jousting or witch trials. We accept the reality of these practices based on written records, letters, and witness reports. They are part of the collective memory of humanity, though none are presently observed.

The same logic extends to events accepted by secular society as historical fact. Later in this book, we will discuss Evolution and the Moon landing—both of which are believed in as firmly established truths, despite not being directly observable or continuously recurring phenomena. Humanity has not witnessed macro-evolutionary transitions within historical time, nor has it returned to the Moon in more than half a century. Yet these are accepted as true on the basis of credible documentation, expert testimony, and consistent reports.

The point is clear: the disappearance of an event from present experience does not undermine the credibility of those who recorded it. If people rationally accept testimony-based knowledge about vanished customs, historical events, or unique achievements, then it is inconsistent to dismiss the testimony for miracles simply because they are not happening before our eyes. The truth of something is not determined by its recurrence, but by the reliability of the witnesses and the coherence of the record they leave behind.

The reliability of Islamic historical records

When discussing miracles within an Islamic framework, the question of evidence inevitably turns to the reliability of Islamic historical records. If the claim is that certain extraordinary events occurred — the splitting of the moon, the water flowing from the Prophet's fingers, the protection of his body from shadow during battle — then the value of these reports depends first and foremost on the credibility of the records that transmit them. It is here that Islam stands in a uniquely strong position. Unlike any other religious or historical tradition, Islam developed a rigorous and systematic method of preserving, verifying, and transmitting information from the very first generation. This is not a matter of theological presumption or insider belief, but a matter of demonstrable, objective method — a discipline that would be considered revolutionary even by modern standards of historical documentation.

From the earliest days, Muslims treated reports about the Prophet ﷺ, his companions, and subsequent generations not as loose traditions or cultural memories, but as verifiable data requiring scrutiny. Every statement, whether about law, history, or miracle, was tied to an identifiable chain of transmission (isnād), each link in which could be investigated for accuracy, character, and continuity. Biographical dictionaries numbering in the tens of thousands were compiled to record the details of every known transmitter — their lifespan, teachers, students, memory, reputation, and even the nature of their mistakes. Through this network, Islamic history was preserved with a level of transparency and traceability unmatched by any other civilization.

The result is that when Muslims speak of miracles, they do not rely on vague traditions or anonymous legends, but on a system of evidence that modern historians can examine, test, and cross-compare. The question, then, is not whether miracles are possible, but whether the reports that describe them meet reasonable standards of reliability. And when placed beside the methods by which we know of other ancient events — whether from Greece, Rome, or early Christianity — the superiority of Islamic record-keeping becomes unmistakably clear.

When assessing the reliability of any historical record, two fundamental questions must be answered. The first is whether the method of transmission itself was reliable — that is, whether there were effective mechanisms in place to ensure accuracy and prevent distortion as information passed from one generation to the next. The second is whether the people involved in that transmission were honest and trustworthy — whether they cared about truth, and whether they could be relied upon to preserve what they heard without fabrication or negligence. In the case of the Islamic tradition, both questions can be addressed not through faith-based assumptions but through tangible, verifiable evidence. As we will see, the system devised by early Muslim scholars ensured a precision in record-keeping unparalleled in human history, and the character and integrity of those who transmitted these records were subjected to levels of scrutiny unknown in any other civilization.

One of the strongest pieces of objective evidence for the reliability of hadith transmission comes from linguistic analysis — specifically, stylometry. Stylometry is a scientific method used to determine authorship by identifying unique linguistic patterns in a text. It has been applied in law and literature to detect forgeries, uncover anonymous authors, and even solve crimes. For example, stylometric analysis confirmed Theodore Kaczynski as the author of the Unabomber Manifesto and helped establish that the disputed play Double Falsehood was written by Shakespeare. Similar linguistic analysis is also used in Biblical studies to determine which letters attributed to Paul were genuinely written by him.

When the same method is applied to the Hadiths of Sahih al-Bukhari, the results are remarkable. Multiple independent studies have shown that the sayings attributed to Prophet Muhammad ﷺ in Sahih al-Bukhari all share a single authorial signature — meaning, linguistically, they are the words of one person. The same studies also found that the Qur'an and the Hadiths have completely distinct linguistic profiles, showing that the Prophet's speech and the divine revelation are clearly separate.

"In this investigation, two visual analytics based clustering approaches have been employed to make a visual authorship clustering of 25 religious text segments.

In the first approach (Hierarchical clustering), the resulting dendrogram shows two separated sharp clusters (the Quran cluster in the right and Hadith cluster in the left). We can see that there is no intersection between the different clusters and that the final linkage is extremely weak since the corresponding distance is very large. This result shows that there are probably two authors: Quran Author and Hadith Author.

In the second approach (Fuzzy C-mean clustering), which is an automatic clustering technique, the resulting 3D representation shows two main clusters: a Quran cluster located at the top right area and Hadith cluster located at the bottom left area of the 3D representation. Although the Quran cluster is more condensed, the two sets of text segments have been automatically organized into 2 sharp clusters (with different symbol markers), showing that there are two main authors: Quran Author and Hadith Author and that the two authors are different."[27]

This finding has major implications. The Hadiths in Sahih al-Bukhari were collected from numerous transmitters, spread across different cities and generations — many of them long before Imam al-Bukhari himself. If the transmitters were fabricating, paraphrasing, or corrupting these reports, the sayings would show stylistic variation typical of multiple authors. Instead, they show one consistent linguistic pattern. The only reasonable explanation is that these sayings were faithfully transmitted from a single historical speaker — Prophet Muhammad ﷺ.

In other words, stylometric analysis confirms what Muslim scholars have maintained for centuries: that the Hadiths in Sahih al-Bukhari are not later inventions or paraphrases, but the accurately preserved words of the Prophet himself. This demonstrates both the honesty of the transmitters and the reliability of the transmission method that preserved their reports across generations.

A common objection to the stylometric argument is that the Hadiths may have been fabricated in one unified linguistic style, meaning that they could appear to have a single author even if they came from multiple fabricators. However, this objection quickly falls apart upon closer scrutiny.

Forgery, like any craft, varies in skill. Most people are poor at it, a few are average, and only a handful reach mastery. For this contention to hold, one would have to assume that numerous transmitters across regions and centuries all became expert forgers — so consistent in their imitation that modern authorship detection tools, sophisticated enough for use in legal cases, could not tell the difference. Such an assumption is implausible to the point of absurdity.

More decisively, stylometric studies have directly tested this hypothesis. Independent researchers have compared authentic (ṣaḥīḥ) and fabricated (mawḍūʿ) Hadiths and found that they are linguistically distinct. One study, for example, used transformer-based models to differentiate between authentic and fabricated Hadiths with a 92% accuracy rate. Here is the reference:

"Muslims rely primarily on the Quran and the Hadiths in all their spiritual life and consider them as sacred sources. If the Quran is God's word, then the Hadiths are God's instructions in the words of the Prophet Muhammad. Since Hadiths are transmitted through multiple narrators, they have been extensively studied to ensure their authenticity.

The purpose of this study is to detect fabricated Hadiths, or Mawdu, which is the type of Hadith most rejected by Muslim scholars. The study utilizes the central text and content of Hadith, Matn, rather than solely focusing on Hadith chain of narrators, Sanad. In order to accomplish this, we create and release the first dataset dedicated to Mawdu Hadiths, called MAHADDAT. Furthermore, we set up a Mawdu Hadith (MH) detection system based on a transformer language model, BERT, achieving a 92.47% F1MH score."[28]

These results are striking. If the Hadiths had all been forged in a single style, both authentic and fabricated collections should display similar linguistic patterns. Instead, they form two clearly distinct stylometric clusters. This confirms two things: first, that authentic Hadiths are not fabrications, and second, that Muslim scholars were remarkably accurate in distinguishing genuine reports from false ones.

In other words, stylometry — an entirely modern, secular tool — independently validates the traditional isnād system. Both arrive at the same conclusion: ṣaḥīḥ reports come from a consistent and truthful source, while mawḍūʿ reports do not. That such agreement exists between classical hadith methodology and contemporary linguistic science is powerful evidence that the transmission system of Islam was not only sincere but objectively reliable.

Among the most compelling proofs for the authenticity of the Hadith and Sirah literature is what historians call the criterion of embarrassment: the idea that reports describing events or details unfavorable to their transmitters are unlikely to have been fabricated. This principle, widely used in secular historiography, provides striking evidence for the reliability of Islamic historical records.

There are two kinds of information in historical sources that strongly suggest authenticity: (1) details that go against the interests or image of the narrators and (2) details that would have been easily altered if tampering were possible. Both are found abundantly in the Hadith and Sirah.

As even non-Muslim military historians have noted,[29] the accounts of early Islamic campaigns display a level of realism and internal consistency that would be impossible if they were forgeries. These records include countless episodes that Muslims would have no reason to invent. Why, for instance, would early Muslims preserve reports of the Prophet ﷺ being caught off guard at Badr, or of defeat at Uhud? Why would they record failed missions or internal complaints against the Prophet's decisions? Fabricators tend to glorify their heroes — yet the Islamic sources preserve both triumph and setback with equal honesty.

The case of ʿAbbās ibn ʿAbd al-Muṭṭalib (RA) is illustrative. He was the Prophet's uncle and the ancestor of the Abbasid dynasty, yet our sources agree that he accepted Islam only after years of opposition — perhaps as late as the conquest of Makkah. Had the Abbasids invented or edited the Sirah, they would surely have made their forefather an early and heroic believer. Similarly, the Umayyad ancestors — Abu Sufyan and Muʿawiyah (RA) — are depicted as late converts, even adversaries before Islam. These are hardly the narratives one would expect from dynasties supposedly inventing their own past.

Other examples multiply this point. Reports describe Muslim mistakes, infighting, and even accidental deaths in battle — such as the tragic killing of Ḥudhayfah ibn al-Yamān's father by Muslim hands at Uhud.

Likewise, the assassination of Kaʿb ibn al-Ashraf is narrated with all its messy realism: three Muslims struggling, one wounded, and the enemy only subdued through desperation. Such accounts are not the polished legends of mythmakers; they are the rough edges of real history.

If Muslim historians were willing to transmit incidents so unflattering — and if such reports carry fully authentic isnads — then the charge that isnads were freely invented collapses. Why would forgers attach verified chains to stories that seem to harm their own image?

The same logic applies to politically sensitive material. The Hadiths contain deep praise for ʿAlī ibn Abī Ṭālib (RA) and his family — narrations the Umayyads and Abbasids had every reason to suppress. Yet these reports not only survived but are found in the most authentic collections such as Sahih al-Bukhari. Likewise, Sīrah Ibn Isḥāq, compiled under the Abbasids, still preserved incidents and virtues that politically favored the Prophet's family, even during active rebellions by the ʿAlid factions.

If the most powerful dynasties could not erase narrations that conflicted with their interests, then it is implausible that they could alter anything else. This is decisive historical evidence — understandable to both scholars and laymen — that the Hadith and Sirah were not politically engineered but faithfully preserved.

Another clear proof of Hadith authenticity lies in the isnād system — the chains of transmission. If these isnāds had been fabricated, we would expect narrations to be attributed to the most senior Companions to make them sound more authoritative. Yet what we actually find is the opposite.

The seven Companions who have narrated the most Hadiths — Abu Hurairah (RA), Abdullah ibn Umar (RA), Anas ibn Malik (RA), Aisha bint Abi Bakr (RA), Abdullah ibn Abbas (RA), Jabir ibn Abdullah

(RA), and Abu Saʿid al-Khudri (RA) — were all among the younger Companions. They were not the earliest converts, nor those who spent the most time with the Prophet ﷺ.

If the isnāds were being fabricated, the fabricators would have naturally chosen senior figures such as Abu Bakr (RA), Umar (RA), or Ibn Masʿud (RA) as sources to give their narrations more weight. The fact that they did not, and that the vast majority of narrations come from younger Companions, strongly indicates that the isnāds were transmitted honestly rather than invented.

This point is reinforced by the case of Abu Hurairah (RA). Despite spending only about three years with the Prophet ﷺ, he narrated more Hadiths than any other Companion — over five thousand. If people were inventing isnāds, they could easily have attributed those Hadiths to Companions who were with the Prophet ﷺ far longer. Yet the historical record preserves reality exactly as it was, even when it appears counterintuitive.

Moreover, the detailed biographies of these Companions have been consistently transmitted and could not have been altered. If historical tampering were possible, figures like Abu Hurairah (RA) could have been presented as early converts to make their narrations more convincing. The fact that even such details remained unchanged shows the integrity of the entire system and the sincerity of those who preserved it.

This pattern within the isnād data itself is powerful evidence that the transmission of Hadith was not a later fabrication but a precise and reliable historical process. Another argument for the reliability of isnāds builds on this and focuses not on who narrated the Hadiths, but on the structure and distribution of authentic isnāds themselves. This argument is distinct from the previous one because it is not based on the criterion of embarrassment but on statistical likelihood.

The isnād system is the backbone of Islamic historical transmission. If isnāds are unreliable, the Hadiths they carry are unreliable; if isnāds are reliable, then the Hadiths themselves stand on solid ground.

Let's compare the two largest Hadith collections in Islam — the most authentic Sunni and Shia sources. The most authoritative Shia Hadith collection, Kitāb al-Kāfi, contains around 16,000 narrations. In contrast, the Sunni Musnad Ahmad includes over 27,000. According to major Shia scholars, including Zayn al-Dīn al-ʿĀmili (ash-Shahīd ath-Thānī), only about **31%** of al-Kāfi's narrations have authentic (ṣaḥīḥ) isnāds, while the rest are classified as weak or doubtful.

By contrast, the majority of Musnad Ahmad's narrations — roughly 75% or more — have authentic isnāds. These assessments are based on each tradition's own scholarly criteria: Shia scholars for Shia Hadiths, Sunni scholars for Sunni Hadiths.

Now, consider the implications. If it were truly easy to fabricate authentic isnāds, as claimed by critics, then both Sunni and Shia compilers should have roughly similar proportions of authentic isnāds. Both groups had strong motivation to legitimize their own traditions. Yet the Shia, despite that motivation, could not fill al-Kāfi with authentic isnāds.

The clear imbalance between the two traditions shows that fabricating authentic isnāds was not easy at all. It required real transmission chains and reliable narrators, not arbitrary invention. Therefore, the prevalence of authentic isnāds in the Sunni Hadith corpus cannot be explained by fabrication — it indicates genuine preservation.

In short, if isnāds could be easily forged, al-Kāfi would have as many authentic chains as the Sunni collections. The fact that it does not proves that fabricating genuine isnāds was extremely difficult, and that the ṣaḥīḥ isnāds found in Sunni Hadith works are, in all likelihood, real and historically grounded.

Another powerful line of evidence for the reliability of Islamic historical records comes from archaeology and early non-Muslim sources. The logic behind this argument is simple: when later written records match early material evidence, that is strong proof of historical preservation.

In the case of Islam, we possess numerous inscriptions and early documents—dating to within decades of the Prophet ﷺ—that confirm details found in our Hadith and Sīrah literature. These include early non-Muslim sources describing the emergence of a Prophet among the Arabs who called to the God of Abraham, forbade idolatry, and taught moral laws. They even mention phrases and ideas found in authentic Hadiths, such as the Prophet ﷺ having "the keys to paradise."

Similarly, early Hijazi inscriptions provide striking corroboration of the historical record. Some mention specific events such as the death of ʿUmar ibn al-Khaṭṭāb, the reign of ʿUthmān ibn ʿAffān, and his later assassination—all recorded with the same dates given in our Islamic sources. Others include expressions like "In the name of God, the Beneficent, the Merciful," confirming that Muslims were using the Basmala in exactly the same form known today.

Inscriptions also mention prominent Companions such as ʿAbd al-Raḥmān ibn ʿAwf, Abū ʿUbaydah ibn al-Jarrāḥ, and Muʿāwiya ibn Abī Sufyān—names and events entirely consistent with our transmitted accounts. The fact that these early inscriptions, written by different people in different regions, align so precisely with the written Hadith and historical reports demonstrates that the Islamic records were not later inventions but faithful preservations of real events and practices.

Taken together, this body of archaeological and textual evidence offers powerful external corroboration for the accuracy and reliability of the Hadith and the broader Islamic historical record.

These are numerous published inscriptions. All of these can be found in The Arabic & Islamic Inscriptions: Examples Of Arabic Epigraphy[1] 1st century sources regarding Prophet Muhammad ﷺ and Islam can be found at Dated And Datable Texts Mentioning Prophet Muhammad From 1-100AH/622-719CE[2] and Dated Muslim Texts From 1-72 AH / 622-691 CE: Documentary Evidence For Early Islam[3] There are many more recently discovered inscriptions but some of these have not been dated. Many of these inscriptions are mentioned by an Orientalist scholar in the following article Early Arabic Rock Inscriptions related to the Qur'an: Juan Cole's List[4] I will give an example of some of these inscriptions and how they prove the reliability of Islamic historical records.

In conclusion, when we look at the historical, textual, and archaeological evidence together, a consistent picture emerges: the transmission methods of Islam were exceptionally rigorous, the transmitters demonstrably honest, and the records independently verified by external sources. The reliability of the isnāds and the preservation of the matn demonstrate a level of historical accuracy unparalleled in any other ancient tradition.

Further details of these and other arguments—as well as responses to common objections raised against Hadith reliability—can be found in my book Sunni Hadiths and Sīrah: Proofs of Historical Reliability.

Conclusion

1. https://www.islamic-awareness.org/history/islam/inscriptions/
2. https://www.islamic-awareness.org/history/islam/inscriptions/earlysaw
3. https://www.islamic-awareness.org/history/islam/inscriptions/earlyislam
4. https://www.juancole.com/quran-commentary-in-context/inscriptions-related-quran

In this chapter, we have examined the positive evidence for miracles by addressing the foundation on which all historical claims rest—the reliability of the records that preserve them. We began by clarifying how testimony and historical transmission can serve as valid forms of evidence, and then demonstrated that, among all historical traditions, the Islamic records stand out for their unparalleled precision and preservation.

Through examining the methods of isnād transmission, the consistency of matn content, the honesty of transmitters, and the external confirmations from archaeology and early sources, it becomes evident that the Hadith and Sīrah literature meet and exceed every reasonable standard of historical reliability. When such a record reports the occurrence of extraordinary events, the burden of proof shifts from the believer to the skeptic.

In the next chapter, we will move from the general to the specific—from the reliability of Islamic historical records to the miraculous events these records preserve. We will examine the evidence for the major Islamic miracles in light of the principles established here, and also consider miracles reported in other traditions to compare their historical credibility.

Miracles: The Historical Record — Islamic and Non-Islamic Case Studies

In the previous chapters, we examined the intellectual foundations for accepting miracles as a part of human history. Chapter One outlined the conceptual framework of miracles, addressing common philosophical objections and establishing why extraordinary events are not, in principle, incompatible with reason. Chapter Two built on this by focusing on the reliability of historical testimony. There, we discussed how historians evaluate eyewitness accounts, why testimony remains a fundamental source of knowledge, and how objections against miracle reports often rely on selective skepticism rather than consistent methodology. That chapter also assessed the historical credibility of Islamic sources, showing that the transmission of reports about the Prophet Muhammad (SAW) rests on a rigorous and uniquely preserved tradition.

This chapter applies those foundations to the specific case of the miracles attributed to the Prophet Muhammad (SAW). It will present the overall evidence for his miracles, address prominent Orientalist claims—especially the assertion that the Qur'an does not affirm his miracles and that Hadith reports are fabrications—and provide detailed discussions of major events such as the splitting of the moon. It will also survey other well-attested prophetic miracles, examine supernatural claims outside Islam, and analyze notable non-Islamic events such as the Fatima apparition. The aim is to evaluate all these reports using consistent principles and to show why the Islamic miracle tradition remains historically and rationally defensible.

The Overall Case for the Miracles of Prophet Muhammad SAW

The foundational case for the miracles of Prophet Muhammad (SAW) is grounded in the remarkable reliability of Islamic historical records.

From the earliest generations of Muslims, scholars developed meticulous methods to preserve accounts of the Prophet's life, teachings, and actions. Every narration was scrutinized through a chain of transmitters, and only reports that met strict criteria of authenticity were accepted. This rigorous approach ensured that the historical record remained remarkably precise, allowing later generations to trace events back to their original sources with confidence.

Within these records, reports of the Prophet's miracles appear repeatedly and consistently. They are documented across a wide range of historical works, spanning different times, regions, and scholarly schools. While not every miracle is transmitted through mass-chains, the recurrence of these reports in independent narrations highlights a persistent recognition of extraordinary events in the Prophet's life.

The repeated documentation of these miracles is significant. It shows that these events were not minor anecdotes or legends added later, but rather central aspects of how the Prophet's life was understood and remembered by his contemporaries and early followers. The sustained attention given to these reports reflects the early Muslim community's view that such extraordinary occurrences were integral to his prophetic mission and worthy of careful preservation.

Some of the miracles of Prophet Muhammad (SAW) are reported as mutawātir, meaning they were transmitted by such a large number of independent narrators that denial becomes virtually impossible. These mass-transmitted reports carry a high degree of certainty and are considered exceptionally strong evidence within Islamic scholarship.

However, the majority of miracles do not have mutawātir chains. This is often cited by critics as a reason to doubt their authenticity. It is important, however, to place this in perspective: most events in any historical record, religious or secular, are not reported by multiple, independent witnesses. Historians routinely rely on single or limited chains of testimony when assessing events, especially those that are extraordinary or rare. The absence of mass transmission, therefore, does not render a report unreliable or historically irrelevant.

In the case of the Prophet's miracles, even non-mutawātir reports often come from chains that are strong, well-documented, and critically examined by scholars. When considered collectively, the repeated and consistent recording of these miracles across multiple sources provides a compelling historical basis for their credibility.

A frequent objection to the miracles of Prophet Muhammad (SAW) is the claim that they lack independent, non-Muslim corroboration and that Muslim transmitters cannot be trusted because of potential bias. Critics argue that, since the reports come from believers, they might be exaggerations or inventions intended to glorify Islam or the Prophet.

This objection, however, applies equally to the vast majority of historical events. Moreover, the potential bias of Muslim transmitters must be weighed against the likely bias of non-Muslims. If a Christian, Zoroastrian, or other contemporary observer had recorded the miracles of the Prophet, they would have been inclined to omit or distort events that favor Islam and challenge their own religious worldview. Therefore, the presence of consistent Muslim testimony, carefully scrutinized through established methods of verification, remains a credible foundation for the historical reality of these miracles.

The Qur'an, the Hadith, and the Reality of Prophetic Miracles

One of the most commonly cited arguments by some Western scholars and critics of Islam is that the Qur'an does not explicitly mention the physical miracles of Prophet Muhammad (SAW), even though it recounts the miraculous deeds of earlier Prophets such as Moses (AS) or Jesus (AS).[30] From this, they argue that the numerous accounts of miraculous events found in the Hadith literature may have been later fabrications created by his followers. While it is true that the Qur'an mentions far fewer miracles of Muhammad (SAW) than it does of earlier Prophets, it does include references to a few miracles and even prophecies of the Prophet, which are often overlooked by these critics.

Some scholars and critics take this argument further, claiming that the Qur'an itself suggests that Muhammad (SAW) did not perform any miracles, or that the Qur'an alone should be regarded as his sole miracle.[31] According to this view, any claim of additional physical miracles would be inconsistent with the Qur'anic text. Such arguments are often repeated in academic and polemical literature, though as we will see later in this chapter, the Qur'an does in fact mention some miracles and prophecies of Prophet Muhammad (SAW), directly countering these claims.

There is a clear reason why some Western scholars and critics emphasize the supposed absence of Muhammad's miracles in the Qur'an. While the Hadith literature contains an enormous amount of material documenting the Prophet's life, many of these scholars do not consider the Hadiths to be fully historically reliable. Because of this, they often treat the Hadith accounts of miracles with skepticism, arguing that they may have been later additions or embellishments by Muslim communities.

The Qur'an, in contrast, is supported by extremely early manuscripts that are widely accepted by modern scholarship as dating back to the time of the Prophet himself. This makes it much harder for critics to dismiss its content, and it provides a solid foundation for the historical credibility of any references it contains. As a result, some critics feel compelled to claim that the Qur'an itself does not indicate that Prophet Muhammad (SAW) performed miracles or made prophecies.

The reasoning behind this argument is that if the Qur'an did record such events, it would serve as an early and authoritative source for them. Acknowledging this would directly challenge their secular worldview and their skepticism toward the Hadiths, which are considered less reliable by these scholars. In their view, the absence of explicit references to miracles in the Qur'an strengthens the argument that the Hadith reports of miracles were later creations.

Yet, the Qur'an does in fact mention a number of miracles and prophecies associated with the Prophet. One notable example is the prophecy in Surah Ar-Rum, which predicts the outcome of a future event — a prophecy that the plain, literal reading clearly attributes to Muhammad (SAW).[32] This directly contradicts the claim that the Qur'an is silent on his miraculous abilities or prophetic foretelling, and it demonstrates that even within the earliest and most reliable textual source of Islam, there is evidence supporting his miraculous and prophetic role.

A simple but important fact is often overlooked in these debates: the Qur'an is not intended to serve as a full biography of Prophet Muhammad (SAW). Its primary purpose is guidance, legal instruction, and spiritual teaching, rather than providing a detailed chronological account of the Prophet's life. Indeed, the name of Prophet Muhammad (SAW) appears only four times throughout the entire Qur'an. By comparison, the name of Prophet Musa (AS) appears over one hundred times, Hazrat Maryam (AS) is mentioned thirty-four times, Prophet Isa (AS) twenty-five times, and Prophet Ibrahim (AS) sixty-nine times. These differences highlight the Qur'an's focus on conveying moral and theological lessons rather than offering an exhaustive historical record of Muhammad's life and deeds.

Given this context, the expectation that all—or even most—of the miracles of Prophet Muhammad (SAW) should be recorded in the

Qur'an is a baseless assumption. The Qur'an provides selective references to events and signs that support its spiritual and ethical objectives. It is not intended to catalog every miraculous occurrence, nor does its relative brevity on Muhammad's life imply the absence of miracles. Critics who base their arguments on the supposed "silence" of the Qur'an misunderstand both the nature of the text and its intended purpose. In reality, the Qur'an offers ample guidance and evidence of the Prophet's prophethood, even if it does not detail every miracle that later sources, such as the Hadith, describe.

Some critics point to a handful of Qur'anic verses to argue that the Qur'an itself denies that the Prophet Muhammad (SAW) performed any miracles. However, with one major exception, these verses are simply the Qur'an quoting the objections and demands of the disbelievers—not endorsing their claims. For instance:

"And they say, 'Why has a sign not been sent down to him from his Lord?' Say, 'Indeed, Allah is Able to send down a sign, but most of them do not know.'" (Sahih International)[33]

This verse does not say the Prophet (SAW) performed no miracles. It only records that some non-Muslims demanded additional signs, just as many people throughout history continued to demand new signs even after witnessing earlier ones. Similarly, in Surah Yūnus, the Qur'an again quotes the objection of the disbelievers:

"And they say, 'Why is a sign not sent down to him from his Lord?' So say, 'The unseen is only for Allah [to administer], so wait; indeed, I am with you among those who wait.'"[34]

Again, this is merely the Qur'an recounting the complaints of those who refused to accept evidence. It is not a statement negating miracles. When these verses are read in context, the argument collapses: the Qur'an often quotes the statements of non-believers as part of its narrative and then refutes or contextualizes them. To take these quoted objections as the Qur'an's own position is simply a category mistake. The Qur'an quoting a demand does not mean the Qur'an affirms what is being demanded or denied.

It is also important to recognize a universal human pattern: people frequently deny evidence that conflicts with their worldview. Critics of Islam today deny the clear prophecies of the Prophet (SAW), even though Muslims can list dozens of them with detailed documentation. Theists often insist that atheists "have no evidence," and atheists say the exact same thing about theists. Christians frequently reject the evidence Muslims present for Islamic miracles, just as Muslims reject miraculous claims presented in other religions. Hindus and Buddhists dispute each other's miracle claims. In every case, one side completely denies the other side's evidence, not because evidence is lacking, but because accepting it would require admitting they are wrong. This universal human tendency shows that mere denial, whether by Quraysh in the 7th century or by critics today, is not evidence that miracles did not occur.

There is one exceptional case in the Qur'an where a specific miraculous sign requested by the Quraysh was not granted. The verse states:

"And nothing has prevented Us from sending signs except that the former peoples denied them. And We gave Thamūd the she-camel as a visible sign, but they wronged her. And We do not send the signs except as a warning."[35]

The traditional explanation is that when a community demands a particular miracle and God fulfills that exact request, rejecting it afterward brings immediate and severe punishment, as occurred with previous nations. Out of mercy for the Prophet's own people, Allah did not send the specific sign they demanded — not because the Prophet ﷺ

lacked miracles, but because responding to their challenge would have sealed their destruction.

This principle is not unique to Islam. The New Testament records a nearly identical episode involving Jesus. When certain scribes and Pharisees demanded a sign from him, he refused, saying:

"Then certain of the scribes and of the Pharisees answered, saying, Master, we would see a sign from thee. But he answered and said unto them, An evil and adulterous generation seeketh after a sign; and there shall no sign be given to it, but the sign of the prophet Jonas."[36]

Christians do not interpret this refusal to mean that Jesus performed no miracles. Rather, they understand it as a rejection of a hostile and insincere demand. If refusal in the Gospels does not imply an absence of miracles, then consistency requires the same conclusion regarding Qur'an 17:59. The Islamic interpretation thus aligns with a broader scriptural pattern: God may withhold a requested sign from a resistant audience, while still granting miracles in other contexts.

In summary, the Qur'an's refusal to grant a specific miracle to the Quraysh does not indicate that Prophet Muhammad ﷺ lacked miraculous ability. Classical Islamic interpretation explains this refusal as a matter of divine wisdom and mercy, ensuring that the people were not destroyed due to their persistent rejection. This principle finds a parallel in the Gospels, where Jesus likewise declined to perform a demanded sign for an insincere audience. Understanding these verses in context shows that the objections of the Quraysh, and the limited references to miracles in the Qur'an, do not undermine the historical reality of the Prophet's miracles. Rather, they highlight a consistent divine pattern in which miracles are granted according to wisdom and the readiness of the audience to accept the truth.

Assessing Alleged Motives for Fabrication

A common objection to the miracles of Prophet Muhammad (SAW) is that Muslims allegedly invented these reports to justify Islam and glorify the Prophet. Critics often present this possibility as sufficient reason to dismiss the miracle reports as historically unreliable. However, this reasoning is deeply flawed and inconsistent. Simply pointing to a potential motive for fabrication cannot, by itself, invalidate historical testimony. If we applied this standard universally, almost all historical accounts—and many modern reports—would be dismissed.

Consider the modern world: governments frequently have motives to cover up crimes, mistakes, or scandals. Yet we do not automatically reject all governmental records or historical accounts of political events simply because officials may have had an interest in shaping public perception. Similarly, the medical industry has strong financial incentives: pharmaceutical companies profit from vaccines and treatments,[37] while media organizations that rely on advertising revenue are influenced by industry funding.[38] If the mere possibility of financial or political motives were sufficient to dismiss testimony, then virtually all reporting on modern medicine, health policies, or public controversies would be invalid.

Even in ancient history, we see similar dynamics. Roman and other imperial historians wrote accounts that glorified their rulers and empires, clearly motivated by politics or prestige. Yet scholars do not reject these accounts outright; instead, they critically evaluate sources, corroborate independent reports, and consider patterns of reliability. In the case of Prophet Muhammad (SAW), the rigorous methods developed by early Muslim scholars for preserving reports—including careful scrutiny of chains of narration and critical evaluation of transmitters—function as safeguards against casual fabrication or embellishment.

The point is clear: potential motive alone is insufficient to dismiss historical testimony. Any historian who applies this criterion selectively—accepting secular histories while rejecting Islamic reports—fails to maintain a consistent standard of evidence. The mere fact that Muslims "might have had a reason" to record miracles cannot outweigh the methodological integrity, consistency, and reliability of the historical sources themselves.

There is another massive problem with this contention from the critics of Islam. The Islamic historical record contains a striking number of accounts that are unflattering, politically sensitive, or even counterproductive to communal pride, which strongly challenges this claim.

One notable example is the extensive documentation of Ali ibn Abi Talib (RA) and his central role in the early Islamic community. The Sirah and Hadith literature contain detailed accounts of Ali's virtues, his close relationship with the Prophet, and his active involvement in key events. Later regimes—particularly the Umayyads and Abbasids—had clear incentives to downplay Ali's influence, given the political conflicts surrounding his legacy. If the reports had been fabricated or selectively edited, one would expect these regimes to alter or remove them entirely. Yet these sensitive details survived across multiple generations and regions, demonstrating that the early historians prioritized accurate transmission over selective glorification.

The historical record also preserves countless other examples that could have reflected poorly on the early Muslim community. Disputes among companions, human errors, and moments of hesitation in battle are meticulously documented. The consistency of these unflattering details indicates that the record was not designed as propaganda, but rather as a faithful preservation of events. As one scholar observes:

"Within the context of starting from the sources largely accepted as genuine within the Islamic movement, I depart from the approaches outlined by Donner by accepting the received collections of hadith literature, particularly regarding the military campaigns and operations. This naturally places my approach in contrast to the Tradition-Critical Approach, which sees much of the hadith literature as late-date forgeries. The reason for my acceptance is both internal and based on the logistical analysis of the events. In this context, forgeries would not have been able to match anecdotal recollections to the realities of the logistics involved.

Besides the aforementioned logistic issues, there are concepts related within the early sources that imply authenticity. For example, why would the resistance in Makkah, and even Madinah, be highlighted rather than glossed over? After all, both cities became the base for the Islamic movement. Would it not make sense, if the material was simply created out of whole cloth, to highlight Muhammad's acceptance and not discuss resistance at all? And as for battles, why would we see Muhammad and his followers endure such troubles and trials? If forged, Muhammad would not have been surprised at Badr, and Uhud would have been transformed into a victory. Makkah would not have been conquered by the Prophet, but he would have instead merely returned to the acclamation of the people. Instances where Muslims were ambushed and defeated would have been glossed over or disappeared altogether. The assassination plot against K'ab bin al-Ashraf would have gone smoothly, with only one man killing him instead of three struggling to do so. Al-Abbas, the Prophet's uncle, would be transformed into a full-fledged believer from the start, not a seeming latecomer to the Prophet's cause. Moreover, any evidence of Muslims complaining about Muhammad's gentleness with al-Abbas would have been excised from the record."[39]

This passage highlights a critical principle: if the early Muslim historians had been fabricating material, they would have shaped the record to emphasize triumphs, gloss over resistance, or remove inconvenient details entirely. Yet the opposite occurred. Resistance, trials, human error, and politically sensitive material were preserved in detail, alongside the miracles. This preservation demonstrates both honesty and methodological rigor, directly challenging the idea that the miracle reports were simply inventions intended to glorify Islam.

In short, the survival of politically sensitive and unflattering accounts—including the extensive coverage of Ali ibn Abi Talib (RA), military setbacks, and internal debates—indicates that the early Muslim historians transmitted their sources faithfully, without systematic fabrication for political or religious gain. This strengthens the credibility of the miracle reports as genuinely historical events, rather than propaganda.

Another line of evidence against the claim that Muslims fabricated miracle reports comes from the meticulous preservation of even minute and potentially controversial biographical details. A particularly illustrative example is Hazrat Abu Huraira (RA). Despite spending only a few years in the company of the Prophet Muhammad (SAW),[40] he narrated over 5,000 Hadith, far more than many companions who spent decades with the Prophet.[41]

If early Muslims had been motivated by glorification, it would have been tempting to exaggerate Abu Huraira's time with the Prophet or inflate his authority to strengthen the credibility of his narrations. Yet the historical record carefully preserves the exact duration of his companionship and includes nuanced details about his narrations and reliability. This transparency demonstrates a commitment to factual accuracy over convenience or religious propaganda.

The preservation of such details is not unique to Abu Huraira. Other companions' virtues, mistakes, and disagreements are also recorded in detail, even when including them could have reflected poorly on the early Muslim community. By maintaining this balance, early Muslim historians created a record that is remarkably resistant to charges of selective fabrication. The fact that Abu Huraira's extensive narrations were faithfully preserved despite his relatively short time with the Prophet underscores the integrity of the transmission process.

In short, the careful documentation of both the quantity of Abu Huraira's narrations and the context of his companionship provides compelling evidence of methodological rigor and honesty. It shows that early Muslim historians prioritized accurate transmission over creating a narrative designed merely to glorify the Prophet or the early Muslim community, which in turn strengthens the credibility of reports about miracles.

A particularly compelling argument against the claim that Muslims fabricated miracle reports comes from events in which early Muslims could have easily hidden, altered, or exaggerated historical material to portray a miracle, yet did not. One clear example is the conclusion of the Ghazwa Ahzab (Battle of the Trench). The Qur'an references this event in the following verse:

"O you who have believed, remember the favor of Allah upon you when armies came to [attack] you and We sent upon them a wind and armies [of angels] you did not see. And ever is Allah, of what you do, Seeing."[42]

While the Qur'an mentions divine intervention—the harsh wind and unseen armies—it makes no mention of the political strategy and clever stratagems that were actually employed during the siege. Historical sources provide detailed accounts of the Prophet

Muhammad's (SAW) strategic use of diplomacy to break the coalition. As recorded:

"With negotiations having come to naught, Muhammad decided to resort to a stratagem of deception to break the coalition. The man chosen for this task was Nu'aym bin Mas'ud bin 'Amr of the Ghatafan tribe. He approached the Prophet and informed him that since becoming a Muslim he had not revealed this to any in his tribe, or to others in the Banu Qurayza or Quraysh. Muhammad dispatched him to these tribes to sow distrust in the coalition, instructing Nu'aym that 'war is deceit.' Nu'aym first went to the Banu Qurayzah.

After allaying any fears and confirming his fealty, he noted that they alone would have to remain in Madinah if the coalition siege was a failure. To prevent the Quraysh and Ghatafan from simply breaking camp and departing, he convinced the Jews that they should ask for hostages to ensure that the others would carry the fight through. The Qurayzah agreed.

Nu'aym then went to the Quraysh and met with Abu Sufyan and his advisors. He convinced them that the Banu Qurayzah had contrived with Muhammad to try and seize some of the notable men of the Quraysh as hostages, only to hand them over to the Prophet to be beheaded. He convinced them not to hand over any hostages, and further convinced them to keep this matter a secret. He then went to the Ghatafan and convinced them to follow the same course of action as that of the Quraysh.

The Quraysh soon sent negotiators to the Banu Qurayzah, indicating that their army was suffering from lack of supplies and therefore needed them to engage the Muslims at once and defeat them. They now asked the Jews to commit themselves, but the Qurayszah demurred, first demanding up to seventy hostages to ensure that the Quraysh and Ghatafan would carry through with their plans. The trap was sprung, and the coalition became distrustful of the Jews, now afraid that they had made a secret alliance with Muhammad.

The very next day, a harsh winter wind raced through the coalition camp, extinguishing their fires and blowing their tents apart. This, coupled with the dwindling supplies and the distrust growing among the coalition members, convinced Abu Sufyan to lift the siege."[43]

The political dimension of the siege is striking. While there was a miracle as well, the Prophet (SAW) also used human strategy, diplomacy, and deception—skills that had no direct miraculous component—to break the coalition. Yet, early Muslims did not bury or obscure this strategy to make the miracle appear even more spectacular. They could have easily omitted all references to Nu'aym's role, the manipulation of the Banu Qurayzah and Quraysh, and the complex negotiations, presenting only divine intervention as the explanation. Instead, these historical details were preserved faithfully, even though reporting them made the Muslims appear reliant on human cunning rather than solely on miraculous aid.

This preservation demonstrates the sincerity and methodological integrity of early Muslim historians. They did not tailor the historical record to maximize glory or exaggerate miracles. The inclusion of politically sensitive and strategically nuanced events, alongside the miraculous elements, provides strong evidence that the historical reports were motivated by accuracy rather than propaganda.

In short, the Ghazwa Ahzab episode shows that Muslims transmitted events honestly, including both divine intervention and human strategy. The decision to preserve the political maneuvers alongside the miracle illustrates the credibility of the historical record and powerfully refutes the Orientalist claim that miracle reports were fabricated to glorify Islam.

Finally, it is important to consider the biases of those who claim that the miracles of Prophet Muhammad (SAW) were fabricated. Critics, especially Orientalists, often approach Islamic sources with predetermined skepticism, which influences their conclusions. Many of these scholars assume that Muslims had a motive to invent extraordinary events, while rarely applying the same standard to other religious or secular histories. This selective skepticism distorts the evaluation of historical evidence.

There is a broader pattern of bias within Orientalist scholarship. Whereas biblical scholars routinely debate, critique, and defend their positions in public forums, Orientalists rarely engage with Muslim scholars in similar ways. Their critiques are often one-sided, and excuses for this lack of engagement abound, yet these explanations fail to justify the imbalance. Moreover, general Orientalist tendencies—such as assuming Islamic sources are late, fabricated, or politically motivated—often overshadow careful, source-based historical analysis.

It is also crucial to note that critics themselves are not free from bias. Many have cultural, ideological, or religious predispositions that influence how they interpret evidence. Just as one cannot dismiss Muslim accounts simply because Muslims might have had motives, one cannot automatically reject reports based on the perceived motives or worldview of their critics. For example, Western historians might be inclined to dismiss accounts of miracles or prophecies that conflict with secular or Christian assumptions about history, but this does not inherently undermine the events themselves.

In short, the Orientalist and critical bias must be factored into any assessment of the miracles. Their selective skepticism, failure to engage in open debate with Islamic scholars, and ideological predispositions highlight the importance of evaluating the Islamic sources on their own methodological merits. When assessed fairly, without bias, the historical record consistently supports the credibility of the reported miracles.

The Moon-Splitting: Evidence and Objections

The splitting of the moon is one of the most significant miracles attributed to Prophet Muhammad ﷺ. Historical accounts indicate that the event occurred when the disbelievers of Makkah demanded a visible sign of his prophethood. In response, the Prophet ﷺ pointed to the moon, and by the will of Allah, it split into two distinct parts, clearly observable to the people present. This extraordinary occurrence served as a divine sign confirming his prophethood. Remarkably, this miracle is mentioned in the plain, literal text of the Qur'an, in Sūrah Al-Qamar:

"The Hour has come near, and the moon has split [in two].

And if they see a sign, they turn away and say, 'Passing magic.'

And they denied and followed their inclinations. But for every matter is a [time of] settlement."[44]

The text explicitly describes the moon splitting as a visible event, leaving little room for allegorical reinterpretation. The Qur'an presents it as a direct miracle, witnessed and rejected by the disbelievers, rather than as a vague metaphor or prophecy.

Even though many Western scholars approach the Hadith literature with skepticism and often reject it as a reliable historical source, the Qur'an itself is widely regarded as a historically reliable text, preserved in an early form that can be traced back to the lifetime of the Prophet ﷺ. Here are a few references for this fact:

"It is true that the earlier consensus of scholarly opinion on the origins of Islam has, since the publication of John Wansbrough's Quranic studies and Patricia Crone and Michael Cook's Hagarism, been shattered, and that various attempts at a new reconstruction of those origins have been put forward.

As a whole, however, the theories of the so-called skeptic or revisionist scholars who, arguing historically, make a radical break with the transmitted picture of Islamic origins, shifting them in both time and place from the seventh to the eighth or ninth century and from the Arabian peninsula to the Fertile Crescent, have by now been discarded, though many of their critical observations remain challenging and still call for investigation.

New findings of Qur'anic text fragments, moreover, can ˉ be adduced to affirm rather than call into question the traditional picture of the Quran as an early fixed text composed of the surahs we have. Nor have scholars trying to deconstruct that image through linguistic arguments succeeded in seriously discrediting the genuineness of the Quran as we know it."[45]

"In any case, textual criticism suggests the standard version is the most faithful representation, among the known codices, of the Quran as recited by the Prophet. This appears at first as a curious coincidence; but on second thought it is not surprising: if anyone had the resources to ensure that a reliable version be chosen, if would have been the caliph; and if anybody had more to lose by botching up the task, again that would have been 'Uthman, whose political legitimacy and efficacy as caliph depended completely on the good will of fellow distinguished associates of the Prophet. The remarkable few and minor skeletal-morphemic differences among the codices Uthman sent to the cities is another indication of the care that was put into the process of standardization."[46]

This widespread academic acceptance of the Qur'an as an early, authentic source poses a serious challenge for critics of Islam. The fact that such a significant miracle is recorded in a text that is both early and historically credible makes it difficult to claim that the event was a later fabrication. Unlike Hadiths, which critics frequently dismiss, the Qur'an offers direct, contemporary evidence of the miracle, embedded in the foundational scripture of Islam. This early attestation significantly strengthens the historical case for the moon-splitting miracle.

There are some common contentions against this miracle that need to be addressed. A common critique of the moon-splitting miracle is that we do not observe any visible cracks or fractures on the moon today — so, the argument goes, the moon could not have actually split. But this objection assumes that a miracle must leave permanent physical damage of the kind we would expect from a natural, physical disruption. That assumption doesn't necessarily apply to a supernatural event.

The moon plays a very important role in Earth's system: its gravitational pull is critical for driving tides, and it also helps stabilize Earth's axial tilt, contributing to long-term climate stability.[47] If the moon were to be catastrophically split by purely natural means, such a violent physical disruption could indeed have had serious consequences for Earth's tides, rotation, and climate.

However, from the Islamic perspective, the moon's splitting was a miraculous act, not a geological or astronomical accident. In a supernatural event, Allah could have allowed the moon to split visibly — so that people saw two parts — but then rejoin or restore its structure, or maintain its integrity in a way that does not disrupt its gravitational relationship with Earth. In other words, the miracle could be externally visible without permanently damaging the moon's physical stability or affecting fine-tuned environmental systems.

Thus, the absence of visible cracks today is not an argument against the moon splitting. Rather, it aligns with the idea that the splitting was not a natural, destructive process, but a controlled divine sign — one that wouldn't necessarily leave lasting geological scars.

A common argument against the moon-splitting miracle is the claim that no independent, non-Muslim sources record the event, which some take as evidence that it did not occur. However, this objection overlooks several important factors that make such an expectation unrealistic.

First, the moon is only visible at night, and different regions of the world experience night at different times. If the moon-splitting occurred in Arabia, vast areas of the globe would not have been able to witness it. Even where it was night, historical habits matter: in the 7th century, many people went to sleep earlier than modern times, reducing the likelihood of observers.[48] Furthermore, those who might have seen it could have dismissed it as an illusion or hallucination. These practical limitations make it understandable that reports outside the immediate region would be scarce.

Second, the miracle likely lasted only a short period, perhaps an hour or two. By contrast, the supernova of 1054, which was exceptionally bright and visible even during the day, was documented mainly in China and Japan, with very few records elsewhere.[49] Historical accounts from other regions are scarce despite the visibility of the event, demonstrating that extraordinary astronomical phenomena are not always widely recorded.[50]

A similar logic applies in evolutionary theory. Evolutionists commonly argue that the absence of transitional fossils is not a problem because such fossils would not necessarily be expected to appear everywhere or in abundance. Yet, as we will discuss later in this book, even according to mainstream paleontologists, the arguments about the scarcity of transitional fossils have faced serious challenges. In other words, the same type of reasoning used to dismiss the lack of external records for the moon-splitting miracle is comparable to reasoning that evolutionists themselves rely on — and which is increasingly questioned within their own field.

Third, critics often argue that powerful, literate civilizations such as the Romans or Persians should have recorded such an extraordinary event. Yet both the Christian Byzantine Empire and the Zoroastrian Sassanian Empire were in conflict with the emerging Muslim community. If observers from these societies had witnessed the moon splitting, they would have had little incentive to preserve a record that validated a rival religion. In this sense, religious or political bias could explain the absence of non-Muslim documentation. Ironically, the same reasoning used to claim that Muslim reports are unreliable due to bias could equally apply to hostile non-Muslim sources.

Taken together, these factors—time zones, short duration, human observation habits, and religious or political bias—explain why independent records outside the Arabian Peninsula are limited. The lack of external documentation does not undermine the historical plausibility of the moon-splitting miracle; rather, it is consistent with the conditions under which the event occurred and the socio-political realities of the time.

A common contention regarding the moon-splitting miracle is the claim that the Qur'an is not describing an event that actually happened during the Prophet's lifetime, but rather is referring to a future occurrence. Critics often base this on the style of speech known as the "prophetic past tense," which is sometimes used in Arabic to describe events that are yet to occur.[51] Many Orientalist scholars invoke this as a reason to reinterpret the Qur'anic account, suggesting that it should not be understood as a historical miracle.

However, this argument does not withstand scrutiny. In the very same third verse of Surah al-Qamar, the Qur'an states that the disbelievers (the Kuffar) rejected the event. The explicit mention of their rejection implies that the event had already occurred, because one cannot reject a future occurrence in this context. This contextual detail directly contradicts the claim that the Qur'an is merely describing a future prophecy rather than a witnessed event. Because of this reason, some Orientalists try to argue that this was an event that happened at that time but it was a natural phenomenon.[52]

Some Orientalists argue that the verse does not refer to a literal splitting of the moon but to a natural phenomenon such as a lunar eclipse.[53] This reinterpretation is meant to deny the occurrence of a miraculous event while preserving the Qur'an's textual integrity. Yet the problem with this explanation is linguistic and conceptual: the Arabic terms used in the Qur'an clearly denote a splitting or cleaving of the moon, not an eclipse or any other astronomical event. In other words, the plain, literal reading of the Qur'an supports the description of a miraculous split rather than a natural lunar phenomenon.

Even if one were to accept, for the sake of argument, that this was a lunar eclipse or that Muslims falsely claimed the moon split, the original Orientalist argument collapses entirely. The claim that the Qur'an does not mention or indicate that Prophet Muhammad (SAW) performed any physical miracles is decisively undermined. Whether the miracle is interpreted literally or figuratively, the Qur'an still affirms that extraordinary signs occurred during the Prophet's mission.

The weakness of the Orientalist arguments becomes even clearer when compared to common skeptical claims in other contexts, such as the debates around evolution or the moon landing. The evidence for the moon-splitting miracle, including the early and reliable textual reference in the Qur'an, is stronger and more direct than the evidence skeptics routinely accept for evolution or the moon landing. Yet Orientalists continue to reject the moon-splitting report, relying on strained reinterpretations and selective reasoning. For any honest and rational observer, it is easy to see that the arguments against the moon-splitting miracle fail miserably when judged by the same standards used to evaluate other historical or scientific claims.

The other miracles of Prophet Muhammad SAW

While the splitting of the moon remains the most well-documented miracle of Prophet Muhammad ﷺ, it is by no means the only miraculous event mentioned in the Qur'an. The Islamic scripture records other extraordinary occurrences in the plain, literal text, demonstrating that miraculous events were an integral part of his prophetic mission. Among the most significant of these is the Isra and Mi'raj — the night journey and ascension of the Prophet ﷺ. This event is clearly referenced in the Qur'an in Surah Al-Isra:

"Exalted is He who took His Servant by night from al-Masjid al-Haram to al-Masjid al-Aqsa, whose surroundings We have blessed, to show him of Our signs. Indeed, He is the Hearing, the Seeing."[54]

This verse presents the journey in straightforward, literal language. It describes a night-time travel from the Sacred Mosque in Makkah to the Farthest Mosque in Jerusalem, followed by an ascension to the heavens, framing the event as a divine sign. Importantly, the Qur'an emphasizes the journey as an extraordinary occurrence, not metaphorical, and directly attributes it to Allah's intervention.

The historical context of this verse is crucial. The Prophet ﷺ faced increasing hostility from the Quraysh, who mocked his message and questioned his prophethood. The Isra and Mi'raj, occurring during this period of intense persecution, served as both a spiritual strengthening for the Prophet and a public demonstration of divine support. Early Muslims, aware of the challenges faced by the Prophet and the fragile state of the emerging Muslim community, would have understood this event as a profound sign of Allah's favor and guidance.

Moreover, the Hadith literature provides additional details that illuminate the significance and nature of this journey. The Prophet ﷺ himself confirmed aspects of the Isra and Mi'raj, describing the stages of the ascension, the encounters with previous prophets, and the divine instructions he received. These accounts, preserved with rigorous chains of transmission, corroborate the Qur'anic reference and highlight the extraordinary nature of this miracle.

By presenting the Isra and Mi'raj in such clear, literal terms, the Qur'an demonstrates that the night journey was not merely allegorical or symbolic. Even many Orientalist scholars acknowledge that Prophet Muhammad ﷺ at least claimed to have undertaken a physical journey. While debate may continue regarding the exact nature of the experience, the plain text of the Qur'an firmly establishes that this extraordinary event was considered a real and significant occurrence within the earliest Islamic community.

The plain, literal language of the Qur'an describing the Isra and Mi'raj clearly presents it as a miraculous event directly associated with Prophet Muhammad ﷺ. This demonstrates that the Qur'an itself attributes extraordinary, physical occurrences to him, alongside the moon-splitting. Consequently, the common Orientalist claim that the Qur'an portrays Prophet Muhammad ﷺ as performing no physical miracles is decisively undermined. By acknowledging both the moon-splitting and the night journey, the Qur'an affirms that miraculous events were indeed part of his prophetic mission, directly countering the argument that all reports of his miracles must come from later Hadith fabrications.

It is important to emphasize, however, that merely showing the Qur'an records the Isra and Mi'raj in plain, literal terms does not, by itself, prove that the event actually occurred or serve as definitive evidence for the Truth of Islam. What it does accomplish, though, is debunk the Orientalist claim that the Qur'an portrays Prophet Muhammad ﷺ as performing no physical miracles. The Qur'an clearly affirms extraordinary events associated with his mission, demonstrating that he was indeed credited with miraculous occurrences during his lifetime.

Beyond this, there are reasons to believe that the Isra and Mi'raj genuinely took place, providing evidence for the truthfulness of the Prophet's claims. One compelling line of reasoning is based on the criterion of dissimilarity: why would Prophet Muhammad ﷺ have invented such an extraordinary event? A journey like this — involving travel to distant lands and ascension to the heavens — would have been difficult for the early Muslim community to accept and could easily have caused doubt or ridicule. In fact, many companions themselves found it challenging to fully believe at first, and the account remains astonishing even to many people today. The improbability of fabricating such a story, given its potential to provoke skepticism rather than reinforce credibility, strongly suggests that it reflects an authentic experience rather than a fictional creation.

There are many other miracles as well. One notable example is the miracle involving the weeping tree. Jabir ibn 'Abdullah reported that the Prophet ﷺ used to lean against a date-palm when delivering sermons. When a pulpit was built, the tree cried like a child; the Prophet ﷺ embraced it and it quieted.[55] Another significant miracle is water flowing from the Prophet's ﷺ fingers. Anas ibn Malik reported that a bowl of water was brought to the Prophet ﷺ, and when he placed his hand in it, water flowed out so that about three hundred people performed ablution.[56] There are also miracles of food multiplication. Jabir ibn 'Abdullah narrated that once the Prophet ﷺ had a small amount of food; after he prayed over it, everyone ate to satisfaction, and there was still enough left to fill their bags.[57]

These miracles, while only recorded in Islamic sources, provide compelling evidence when considered alongside the rigorous preservation and transmission of Hadiths. The historical methods employed by early Muslim scholars, the integrity and honesty of the companions, and the critical scrutiny applied to every narration make these accounts powerful testimony for the Prophet's ﷺ extraordinary deeds. They demonstrate that the Hadiths are reliable sources for reporting physical miracles of Prophet Muhammad ﷺ.

Supernatural Reports Outside Islam

Muslims do not only believe in the miracles of Prophet Muhammad ﷺ. Islamic teachings recognize the miracles of earlier Prophets, and many, though not all, of the miraculous events mentioned in the Bible or pre-Islamic texts can be accepted within an Islamic framework. This recognition does not compromise the unique status of Prophet Muhammad ﷺ or the authenticity of Islam; rather, it reflects the broader principle that God has, throughout history, manifested extraordinary signs to guide humanity.

There are also a few reports of miracle-like or supernatural events outside the Islamic tradition that appear to have relatively strong evidence, particularly within Christianity. Some Muslims, in an attempt to engage critics, repeat Western skeptical arguments that attempt to dismiss these events. However, these arguments are often weak and rely on double standards. Just because an event is considered supernatural does not automatically mean it originates from God. Islamic thought allows for the possibility that such occurrences may result from the actions of jinn or other non-divine agents. Accepting the supernatural nature of an event does not obligate a Muslim to credit it to divine revelation or to see it as validating any non-Islamic religion.

Importantly, the existence of supernatural events outside Islam does not prove the truth of Christianity or any other religion. Even Christian and other non-Muslim traditions recognize the possibility of demons or other supernatural agents performing extraordinary acts. In Islam, this perspective is consistent and non-contradictory: extraordinary phenomena may occur in the world without being signs from God.

What distinguishes Islam from other religions, particularly Christianity, is its internal and external consistency. Unlike Christianity, Islam contains no internal contradictions, and it is free from false prophecies. While miraculous events may be reported in multiple traditions, the coherence of Islam, the reliability of its historical record, and the absence of doctrinal errors or failed prophecies underscore its unique authenticity. A Muslim can acknowledge the possibility of supernatural occurrences outside Islam without compromising the truth of Islamic teachings.

The Fatima Apparition: Analysis

Between May 13 and October 13, 1917, three shepherd children in Fátima, Portugal — Lucia dos Santos and her cousins Francisco and Jacinta Marto — reported seeing a luminous "Lady" who identified herself as the Lady of the Rosary. She urged daily prayer and warned of sin, while promising a public sign on October 13 so that everyone would believe. On that date, tens of thousands of people reportedly witnessed a remarkable solar phenomenon, now called the "Miracle of the Sun." Witnesses described the sun appearing as a dull, silvery disc, spinning, changing colors, and moving erratically in the sky, while many who had been soaked in a rainstorm found their clothes drying almost instantly. Other effects reported include the stilling of tree leaves and some physical healings or conversions among the crowd.

The evidence for this event is considered strong because of the large and diverse number of witnesses, the accurate prediction of the date of the phenomenon, and the physical effects that accompanied it. Reports came from people of different social and religious backgrounds, including skeptics and journalists. The consistency of accounts regarding the extraordinary movement and behavior of the sun, combined with the scale of the event, make it difficult to dismiss as a simple mass hallucination or optical illusion. While mass hysteria can produce shared anxiety or psychosomatic effects, there is no evidence that tens of thousands simultaneously experienced the same hallucination, which makes this explanation inadequate. Variations in individual descriptions are normal given the emotional intensity and size of the crowd.

Common objections include suggestions that the event was a natural atmospheric phenomenon or illusion. These explanations fail to account for the large number of independent witnesses, the prior prediction of the date, and the physical effects reported, such as the rapid drying of clothes. Some skeptics also question the children's credibility, but long-term observation of their behavior and testimonies indicates no evidence of deceit or fabrication. Importantly, this could have been a purely supernatural event not related to Christianity; Christians may have interpreted it according to their expectations, but that does not diminish the fact that something extraordinary and miraculous occurred. Muslims can accept the event as a genuine supernatural occurrence without assuming it validates Christian doctrine, as Islamic thought allows for supernatural events originating from God, angels, or jinn.

Conclusion

Taken together, the evidence demonstrates that the miracles of Prophet Muhammad ﷺ rest on a uniquely strong historical foundation. Islamic records are far more reliably preserved than those of any other ancient tradition, and while not every miracle is mutawātir, this is completely normal for historical events. The common claim that the Qur'an denies his miracles collapses once the Qur'an is read in context: it is not a biography, it rarely mentions details of the Prophet's life, and yet it still explicitly refers to both prophecies and physical miracles, including the splitting of the moon. This alone undermines the main Orientalist argument.

Appeals to fabrication also fail. The Islamic tradition openly preserves countless reports that would have been removed if Muslims were manipulating their history. The honesty, consistency, and structure of transmission—combined with the lack of any political or theological motive to invent many of these miracles—makes large-scale fabrication implausible.

The moon-splitting stands out with especially strong evidence in both the Qur'an and hadith. The objections commonly raised—lack of independent reports, supposed cracks on the moon, or claims of a future-tense reading—do not withstand scrutiny. Historical, linguistic, and observational realities all favor the traditional Islamic interpretation. Other miracles, such as al-Isrā' wal-Mi'rāj, food multiplication, water miracles, and the obedience of trees, further reinforce the broader pattern found in our early sources.

Even supernatural events outside Islam do not undermine the Islamic case. Their existence is fully compatible with Islamic theology, as such occurrences may be the actions of jinn rather than evidence for other religions. Islam distinguishes itself not by denying all other supernatural claims, but by presenting a coherent scripture, a consistent theological system, and a prophetic history free from contradiction or failed prophecy.

In summary, when the arguments, sources, and counterarguments are examined fairly, the miraculous signs of Prophet Muhammad ﷺ emerge not as legends or embellishments, but as historically credible events that strongly support the truth of Islam.

Miracles: The Objections — skepticism, naturalism, and their refutation

Throughout history, miracles have stood as one of the most direct challenges to materialist and naturalistic worldviews. They claim to be events that cannot be explained by known physical laws — moments in which divine power intervenes in the observable world. Yet from the Enlightenment onward, a persistent current of skepticism has arisen, denying either the possibility or the credibility of such events. Modern critics often present their case as a defense of reason and science, but their objections, upon closer inspection, often rest on philosophical presuppositions rather than empirical evidence.

This chapter examines the main contentions raised against miracles — both classical and modern — and exposes the logical, philosophical, and historical weaknesses behind them. We will analyze the claim that miracles are inherently improbable, address David Hume's famous argument that no testimony can establish a miracle, and discuss what may be called the "camera problem" — the demand for modern photographic evidence as a condition for belief. Further sections will consider the objection that miracles are not testable or verifiable by scientific means, the allegation that religious believers are motivated by bias or fabrication, and the challenge posed by competing miracle claims across different faiths.

The aim of this chapter is not only to refute these objections but also to demonstrate that belief in miracles can be fully rational when evaluated within a consistent epistemological framework — one that recognizes both the limits of naturalism and the validity of historical evidence.

Inherent improbability of miracles

One of the most frequently repeated objections to miracles is that they are, by definition, the least likely events to occur. It is often claimed that because miracles are "highly improbable," it is more rational to assume that they never happen. This argument, however, is based on a fundamental confusion about the nature of probability. Nothing is intrinsically probable or improbable in isolation — probability is always relative to one's background assumptions and explanatory framework.

For instance, in the discussion on the fine-tuning of the universe, skeptics often dismiss the argument by asserting that the physical constants are not "improbable" at all — they are simply necessary or could not have been otherwise. In other words, they conveniently remove the element of chance by appealing to some unknown law or metaphysical necessity. Yet in other contexts, such as when discussing the origin of life or evolutionary processes, the same skeptics insist that pure chance and unguided natural processes are sufficient to explain extraordinary complexity and order. When the evidence points toward design, they invoke "necessity" to avoid the conclusion; when it points toward purpose, they appeal to "chance" to avoid the same. This selective shifting between chance and necessity is not scientific rigor — it is philosophical convenience.

Similarly, in the case of evolution, the similarities among living organisms are considered improbable only if we assume they arose by random coincidence. If one instead assumes that these patterns result from physical or biological necessity, the improbability disappears entirely. Once again, the choice of framework determines what counts as "improbable." Yet, when the same logic is applied to miracles, skeptics suddenly forget this conditional nature of probability. They act as if improbability were an objective feature of events themselves, rather than a reflection of the assumptions we bring to them.

This exposes a striking double standard. When defending naturalistic explanations, skeptics freely redefine their assumptions to preserve naturalism — invoking necessity when chance seems inadequate, and chance when necessity fails. But when theists point out that miracles are probable given the existence of God, skeptics demand proof of God before allowing that premise into consideration. They shift the burden of proof entirely onto the believer while exempting their own worldview from the same scrutiny.

The reality is simple: if one assumes naturalism — that only material causes exist — then miracles are of course improbable or even impossible. But if one assumes theism — that God exists and may intervene in His creation — then miracles are not improbable at all; they are coherent and even expected under certain conditions. The claim that miracles are "inherently improbable" therefore carries no weight. It is not a statement of science or reason, but an expression of philosophical bias disguised as rational skepticism.

Hume's argument

David Hume's argument against miracles is often presented as a hallmark of rational philosophy, yet it rests on self-contradictory and selectively skeptical assumptions. In his essay On Miracles, Hume claimed that it is never rational to believe a miracle report because miracles, by definition, contradict the "uniform experience" of mankind. The supposed "laws of nature," he argued, are confirmed by this uniform experience, and therefore, any report of a violation of these laws is always less credible than the possibility of error, deceit, or delusion in testimony. [58]

The flaw in this reasoning is immediately apparent. Hume's so-called "uniform experience" is not something he can actually access or verify. No individual can know what all of humanity has experienced; our only access to the experiences of others is through their testimony.

Thus, the very concept of "collective experience" depends on the acceptance of testimony as a source of knowledge. And here, Hume's argument collapses under its own weight.

As Professor C. A. J. Coady points out in his classic work Testimony: A Philosophical Discussion, Hume's approach reflects a reductionist model of knowledge — that is, the idea that testimony is only valid when confirmed by perception, memory, or induction.[59] Yet this model creates a vicious circle: we can only know what others have observed through testimony, meaning that collective experience cannot be used to discredit testimony without presupposing the very thing being denied. Hume depends on testimony to claim that miracles never occur, while simultaneously disqualifying testimony when it affirms that miracles do occur.

And herein lies the most glaring hypocrisy: if one actually considers the collective testimony of humanity, it is overwhelmingly filled with reports of miracles and extraordinary events. From ancient civilizations to modern times, across all continents, cultures, and religions, human beings have consistently reported witnessing occurrences that defy ordinary explanation — healings, prophecies, answered prayers, and phenomena beyond natural causation. The historical record is saturated with such accounts. If Hume truly believed that belief should follow collective human testimony, he would be compelled to accept that the vast majority of humankind throughout history has affirmed the reality of miracles.

But instead, he cherry-picks what counts as "collective experience." Testimony to ordinary events is accepted as reliable; testimony to extraordinary events is dismissed as delusion — not because of evidence, but because of an a priori commitment to naturalism. This is not rational philosophy; it is selective skepticism. Hume's so-called empiricism functions only when it serves his materialist assumptions.

When human testimony supports a naturalistic worldview, it is trustworthy; when it points to the divine, it is suddenly suspect.

In short, Hume's critique of miracles not only collapses philosophically — it collapses historically. The collective testimony of humankind directly contradicts his claim. His "uniform experience" is not the record of mankind but the selective filter of his own worldview. The persistence of this circular argument in Western academia is therefore not a sign of its strength but of the deep ideological bias that continues to masquerade as "reason" in secular thought.

The camera problem

One of the most common modern objections to miracles is the so-called camera problem: "If miracles were real, why don't we see them on camera?" The assumption here is that if divine acts truly occurred, they should be regularly recorded, broadcast, and verified through modern technology. This argument reflects both a misunderstanding of what miracles are and a selective, self-serving view of evidence.

Miracles, by their very definition, are extraordinary events — rare, purposeful interventions that occur according to divine wisdom, not human demand. Expecting them to appear constantly and conveniently for our devices misunderstands their very nature. They are not spectacles for scientific testing but signs for those meant to witness them. To demand constant video proof of miracles is as absurd as demanding a solar eclipse every morning to believe in astronomy.

Moreover, this insistence on visual evidence betrays a shallow understanding of knowledge itself. Human history, science, and daily life depend on testimony and inference, not on camera footage. We believe in historical events, scientific processes, and even the existence of distant galaxies without direct visual verification. The camera is not the ultimate judge of truth — it is merely a tool, and an unreliable one at that, easily manipulated or deceived.

The irony becomes clearer when we apply this same "camera standard" to other beliefs that secularists take for granted. Consider Evolution or the Moon landing. Neither of these phenomena has been directly observed in the way skeptics demand miracles to be. Evolution is not something anyone has ever filmed in action — it is inferred from partial evidence interpreted through theoretical frameworks. Likewise, the Moon landing, though supported by certain forms of evidence, ultimately rests on limited footage and testimony — the same categories of evidence skeptics dismiss when it comes to religious claims.

When confronted with gaps or inconsistencies in these secular narratives, their defenders often resort to ad hoc explanations. For example, when the fossil record lacks transitional forms, the response is, "They'll be discovered later." When inconsistencies in Moon landing footage are raised, explanations such as "camera malfunction" or "necessary reenactment for clarity" are given. These are not fundamentally different from the explanations theists give when asked why every miracle is not publicly visible: that God acts according to wisdom, not spectacle; that miracles are not meant for coercive proof but for guidance.

The hypocrisy lies not in having explanations, but in pretending that only religious believers rely on them. Secularists condemn theists for offering theological reasoning while employing the same type of reasoning themselves — except theirs is wrapped in the language of "science" or "methodology."

Finally, there is the issue of faked footage. In an age of digital manipulation and deepfakes, the idea that "if it's on camera, it's true" is laughable. Even the Moon landing footage — once untouchable —

has been scrutinized for signs of staging or enhancement, a reminder that visual evidence can be produced, edited, or forged. If footage can be faked, then the absence of footage cannot disprove an event, nor can its presence definitively prove one.

In the end, the "camera problem" is not a problem for miracles but a problem for modern skepticism. It exposes an irrational double standard — a world where believers must produce live-streamed divine acts to be believed, while secularists demand blind faith in unobserved evolutionary transitions and government footage from half a century ago. The real issue is not evidence, but arrogance: the refusal to admit that human perception, technology, and interpretation are limited, and that the truth may transcend what fits within a frame.

Lack of testability of verifiability

Another common objection raised against miracles is that they are untestable or unfalsifiable. Skeptics argue that since miracles cannot be repeated under controlled conditions, they fall outside the scope of scientific verification and therefore cannot be considered credible. This claim, however, rests on a fundamental misunderstanding of both science and history — and once again, it exposes a striking double standard in how secular thinkers apply their own principles.

Not everything that is true is scientifically testable or repeatable. Historical events cannot be recreated in a laboratory, yet their reality is not doubted by any reasonable person. The fall of empires, the signing of treaties, or the lives of historical figures are known through testimony, records, and evidence — not through experiment. Miracles fall into this same category: they are singular, historical occurrences, whose credibility depends on testimony and historical verification, not laboratory repetition.

If we were to insist that only testable events deserve belief, then we would have to reject much of what we know, including many things secularists themselves accept as fact. Consider the Moon landings. There were multiple alleged missions to the Moon between 1969 and 1972, yet after that brief period, space travel beyond low Earth orbit abruptly stopped — and for over half a century, no human being has gone beyond that point again. This raises a serious question: if space travel to the Moon were truly straightforward and empirically verifiable, why did it happen only within a narrow window of Cold War competition and then cease entirely, despite massive technological advancements since? The very event that supposedly demonstrates humanity's greatest scientific achievement remains unrepeatable in practice, and verification still depends on official reports and video footage from decades ago.

The same applies to Evolution. It is widely described as a scientific "fact," yet the processes it describes — species gradually transforming into others over vast timescales — are neither directly observable nor experimentally reproducible. They are inferred from fragmentary evidence and theoretical models. As far as falsifiability is concerned, Evolution is one of the least falsifiable frameworks imaginable: any contradictory evidence is explained away through ad hoc reasoning. Missing fossils are said to be undiscovered. Unexpected genetic overlaps are attributed to "convergent evolution." Sudden appearances of new forms in the fossil record are said to be due to "punctuated equilibrium." In every case, the theory shifts to absorb the anomaly rather than face refutation.

Thus, when skeptics dismiss miracles as untestable, they are condemning their own epistemology. The Moon landing is accepted though unrepeatable; Evolution is upheld though unfalsifiable; yet miracles are rejected for the very same reasons. This is not science — it is selective belief. If testability and falsifiability are to be the ultimate standards of truth, then consistency demands that these secular claims be treated with the same skepticism applied to miracles.

Religious bias and motivation for fabrication

Religious bias and motivation for fabrication are real concerns, but they are not unique to religion; political, institutional, and ideological incentives produce distortion and selective reporting across secular domains as well. The U.S. effort to reach the Moon illustrates this clearly: the program was driven as much by Cold War politics and prestige as by pure scientific curiosity—John F. Kennedy repeatedly framed the lunar effort as a race to beat the Soviets and secure geopolitical advantage.[60] Early Cold War planning even contemplated using spectacular, propagandistic displays on the Moon to influence world opinion, as shown by the declassified proposal known as Project A119, a 1958 U.S. Air Force study that considered detonating a nuclear device on the lunar surface for its publicity value.[61] The political context and propaganda logic surrounding the Apollo program created powerful incentives to control the narrative and to prioritize demonstrable national prestige over disinterested verification.[62]

Secular history likewise shows how victors and powerful institutions shape what counts as authoritative accounts. Modern critiques of Western historiography—most famously Edward Said's Orientalism—document how colonial and imperial power produced scholarly traditions that represented non-Western peoples in ways convenient to imperial interests, marginalizing indigenous voices and recasting religious or cultural practices as primitive or "mythic."[63] Concrete examples include Eurocentric narratives that downplay sophisticated Islamic scientific and legal traditions or recast colonial conquests as "civilizing missions," choices that reflect power and ideology as much as archival truth.[64] In short, the historian's archive and interpretive frameworks are often structured by who controlled the press, the archives, and the universities—which means "secular" histories can conceal as much as they reveal.[65]

In scientific and public-health arenas, political and financial incentives create equally strong motivations for selective reporting or narrative control. The MMR-autism scare began with a small Lancet article whose methodological failures and undeclared conflicts of interest were later exposed and the paper retracted—an episode that produced real-world harm and revealed how poor scholarship and questionable incentives can propagate widely before being corrected.[66] More recently, institutional controversies over vaccine policy and data management have led to high-profile resignations that underscore political interference with scientific communication.[67]

Academic incentives and ideological commitments likewise affect evolutionary biology. The Sternberg peer-review controversy, in which a contentious intelligent-design-related article was published and then generated severe institutional backlash, is often cited as an example of how editorial and professional gatekeeping can be used to police orthodoxy, and how the boundary between scientific adjudication and ideological enforcement can blur.[68] Whether one accepts Sternberg's claims or not, the episode shows that debates over evolutionary theory are not purely technical; they are entangled with broader cultural and institutional pressures that can discourage heterodox views.

Finally, the combination of political urgency, institutional control, and limited independent verification helps explain why some secular narratives become entrenched even when the original evidentiary base was narrow. The Moon landings, while supported by substantial technical documentation and testimony, were publicized within a narrow Cold War window (1969–1972) and have not been repeated with humans since—a historical fact that invites reasonable questions about the interplay of politics, technology, and narrative management rather than a rush to dismiss any questioning as irrational.[69]

All of these examples show the same pattern: incentives to promote a triumphant secular story (political prestige, institutional authority, ideological consistency, or financial interests) are pervasive and can produce selective presentation, suppression, or politicization of evidence. This does not prove that particular secular claims are false, but it does mean the critic who points to religious bias must apply the same scrutiny to secular and institutional contexts—and accept that motives for fabrication and narrative control run in every direction.

The problem of genuine cases of fabrication

A common objection raised by skeptics is that there have been instances of fabricated miracles or false claims, and therefore all miracle reports are unreliable. This reasoning, however, is logically fallacious. The existence of some forgeries or exaggerations within a category does not entail that every claim within that category is false. To reject all miracle testimony because a few have been fabricated is no different from rejecting all scientific or historical claims because some have turned out to be fraudulent.

Throughout secular history, fabrication and error have occurred across all fields of knowledge. The discipline of evolutionary biology, for instance, has witnessed notorious examples such as the Piltdown Man hoax, in which an alleged "missing link" between ape and man was presented to the public for over forty years before being exposed as a deliberate forgery.[70] Yet no one argues that the entire field of evolutionary science should be dismissed because of this deception. Likewise, the Haeckelian embryo drawings, used in textbooks for over a century to promote the concept of recapitulation, were later revealed to be deliberately exaggerated.[71] Even today, certain grand claims of evolutionary theory—such as confident predictions about the "junk DNA" being useless or specific transitional fossils being permanent proofs—have been quietly revised or abandoned without any loss of faith in the broader theory.[72]

The same double standard appears in secular historiography. Historians have uncovered countless examples of forged documents, manipulated records, and politically motivated revisions. Medieval Europe produced fabrications such as the Donation of Constantine, once used by the Church to justify papal authority, later exposed as a forgery by humanist scholars.[73] In modern times, entire historical narratives have been reshaped by propaganda or ideology—from Stalin's photographic erasures of political rivals to colonial rewriting of conquered histories.[74] Yet few secular historians would suggest that all of history therefore collapses into myth.

Even space exploration, often portrayed as a bastion of modern scientific certainty, is not immune to similar problems. While the Apollo missions are broadly accepted as genuine, skepticism surrounding the Moon landings is fueled in part by legitimate inconsistencies, photographic anomalies, and the suspicious discontinuation of manned lunar travel after 1972 despite dramatic advances in technology.[75] Whether or not one accepts conspiracy theories, the central point remains: the mere existence of fabrication, propaganda, or questionable evidence in one domain does not justify universal dismissal of all testimony in that domain.

The double standard becomes glaring when one observes how skeptics demand that every miracle claim be rejected if even a single case of fraud is uncovered, while continuing to uphold scientific, political, or historical frameworks that have also produced well-documented falsehoods. If consistency were applied, secularism would have to doubt its own narratives just as harshly as it doubts religious testimony. But it rarely does. Instead, a selective skepticism is maintained—one that treats the occasional falsehood in religion as evidence of inherent unreliability, while treating falsehoods in secular science as innocent growing pains in the pursuit of truth.

This inconsistency reveals that the objection to miracles is not really about evidence but about worldview. When secular narratives are challenged by instances of fabrication, they are reinterpreted, revised, or explained away; when religious narratives face the same, they are summarily dismissed. The believer, however, asks only for consistency: if human error, bias, or fraud does not invalidate the whole of science or history, neither should it invalidate the whole of miracle testimony. Genuine inquiry demands that each claim be weighed on its own merits, not condemned by association.

Competing claims of miracles

Skeptics often argue that since different religions claim miracles that support contradictory beliefs, this invalidates all such claims. If miracles are said to prove the truth of Christianity, Hinduism, or Islam, and these faiths make mutually exclusive theological claims, then—so the argument goes—they cannot all be true. Therefore, miracles, it is claimed, cannot be a reliable indicator of truth.

However, this objection fails to address the actual question of whether miracles occur. At most, it shows that miracles alone do not establish which interpretation of them is correct, not that no miracles happen at all. The existence of counterfeit currency does not disprove the existence of real money; it merely demands a reliable standard of authentication. Likewise, competing miracle claims call for discernment, not denial.

In this respect, Islam holds a uniquely coherent and principled position. It does not deny the miracles of other prophets or previous faiths but explicitly affirms them as part of a continuous divine plan. The Qur'an recognizes the miracles of Moses—such as the staff turning into a serpent and the parting of the sea—as true divine acts witnessed by his people.[76] It likewise confirms the miracles of Jesus, including healing the blind and lepers and raising the dead by God's permission.[77] These are not denied or reinterpreted allegorically but accepted as genuine, historical divine interventions. Islam thus accepts the occurrence of many miracles claimed by other religions while explaining their theological context correctly: that all prophets were sent with the same essential message of tawḥīd (divine oneness) and submission to God.

At the same time, Islam acknowledges the possibility of supernatural occurrences that do not stem from divine sanction. Acts of the jinn, sorcery, or demonic deception can produce extraordinary phenomena that may appear miraculous but arise from other unseen forces.[78] This framework allows Muslims to rationally explain alleged miracles in non-Islamic contexts—some may stem from divine truth, others from deceptive supernatural sources, and still others from human fabrication or exaggeration. Christianity and secularism, by contrast, offer more rigid binaries: the former often regards all non-Christian supernatural claims as demonic, while the latter rejects the supernatural altogether. Islam, however, provides a balanced and consistent middle ground that both affirms the possibility of miracles and distinguishes their sources through revelation.

If one applied the skeptic's "mutually exclusive claims" argument consistently, it would undermine many secular beliefs as well. The theory of evolution, for instance, contains multiple internal contradictions and rival interpretations—Darwinian gradualism, punctuated equilibrium, neutral evolution—each of which challenges the others' explanatory power.[79] Yet secularists do not discard evolutionary theory altogether; they simply acknowledge competing interpretations. Similarly, the numerous and conflicting narratives surrounding the Moon landing—ranging from official reports to persistent counterclaims and inconsistencies—have not led to a wholesale rejection of space travel.[80]

In short, conflicting miracle claims among religions do not refute the possibility of miracles any more than conflicting scientific models refute the existence of the natural world. Islam's position uniquely resolves this issue: it affirms the truth of genuine divine miracles across prophetic history, recognizes the possibility of deceptive or non-divine supernatural acts, and integrates all authentic miracles into one coherent, monotheistic framework.

God of the gaps

The so-called "God of the gaps" objection is among the most common accusations leveled against belief in miracles. The claim goes that whenever the believer encounters something unexplained, he inserts God into the gap in scientific knowledge as a convenient placeholder — and that as science advances, these "gaps" will gradually close, leaving no room for divine intervention. On the surface, this sounds clever; in reality, it is little more than a rhetorical trick that distorts the actual reasoning behind belief in miracles.

No serious theist or theologian argues for God's existence from temporary ignorance. The recognition of miracles does not arise from "we don't know how this happened," but from "we know enough to recognize that this violates the normal order of causes." When an event directly contradicts the established regularities of nature — such as an instantaneous healing or the dead returning to life — the inference to a supernatural cause is not based on ignorance but on knowledge of the natural order and its limits. One must know what nature does to recognize when something happens that nature cannot do.

The irony, however, is that modern materialist arguments routinely commit the very same fallacy they accuse theists of. Consider the history of "Junk DNA." When scientists first discovered large portions of non-coding DNA, they assumed, based on ignorance, that it was meaningless leftover material from evolution — "junk." Yet decades later, research revealed that much of this so-called junk DNA plays crucial regulatory and structural roles in the genome, governing gene expression and maintaining chromosomal stability.[81] The same can be said of "vestigial organs." In the 19th century, evolutionists claimed that organs like the appendix, tonsils, and coccyx were useless remnants of our evolutionary past. Modern biology has shown that these organs serve clear functions — the appendix plays an important role in immune regulation and gut flora maintenance; the tonsils form part of the lymphatic system's first line of defense; and even the coccyx aids in muscle attachment and posture.[82]

In both cases, the argument followed the exact same pattern as the "God of the gaps" fallacy — we do not know the purpose of this, therefore it has no purpose, therefore it was not designed. This is not science but a statement of dogmatic faith in purposelessness. It mirrors the same reasoning that skeptics condemn when used by theists: ignorance elevated to explanation. The difference is that when atheists make the argument, it is treated as "methodological naturalism," and when theists do, it is derided as superstition.

Furthermore, the claim that certain biological structures "had a function but lost it" or "once were useful but are now redundant" only deepens the irony. In designed systems, it is perfectly normal for some components to lack a current function or to have a function that changes over time. A building, for example, can have empty rooms, disused storage spaces, or corridors built for future use — all without ceasing to be a designed structure. Engineers, architects, and software designers routinely include redundancies, backup systems, or multi-purpose components that may not always be in active use. To claim that a structure's lack of visible function proves an absence of design would be absurd in every other context — yet this is precisely what evolutionists argue when they assume that unused DNA or vestigial organs prove randomness rather than purpose. Even by their own reasoning, these features could simply have shifted or latent functions, just as in any intelligently organized system.

Moreover, the "God of the gaps" criticism fails to distinguish between inference to the best explanation and appeal to ignorance. A miracle claim, if well-attested, is not merely an "unknown cause" but an event with identifiable signs of intentionality, timing, and meaning — elements that cannot be accounted for by random physical processes. The believer is not filling a gap but recognizing agency where agency is manifest. By contrast, the materialist "gap" reasoning behind things like junk DNA assumes in advance that there can be no purpose, then interprets ignorance as evidence for purposelessness — a textbook case of circular reasoning.

This double standard once again reveals that the issue is not intellectual but ideological. It is not that theists insert God where science has no answer, but that secularists insert "chance" or "natural necessity" wherever purpose is uncomfortable. In both cases, a metaphysical assumption fills the gap — the only difference is which assumption one prefers.

Mass Hallucination/Hysteria

Among the more desperate attempts to explain away miracles is the claim that they are the result of "mass hallucination" or "mass hysteria."

Skeptics suggest that when large groups of people report witnessing the same miraculous event, they are not describing something that actually occurred, but are collectively deluded or psychologically influenced. On closer examination, however, this idea collapses both scientifically and logically.

The notion of a "mass hallucination" is not recognized in clinical psychology. Hallucinations are defined as perception-like experiences that occur without an external stimulus, and they are always individual phenomena. There are no verified cases in the psychiatric or neurological literature of multiple independent people simultaneously experiencing the same detailed hallucination in the absence of an external cause.[83] Leading researchers Robert Bartholomew and Simon Wessely emphasize that there are no scientifically verified instances of multiple people sharing the same hallucination, and the statistical probability of such an event occurring independently is effectively zero.[84]

What does exist is mass psychogenic illness, formerly called "mass hysteria." This is a real, documented phenomenon, but it has nothing to do with shared visual experiences. Mass psychogenic illness occurs when groups of people develop similar physical symptoms or behaviors through social contagion, often triggered by stress or anxiety. In such cases, people influence each other emotionally or behaviorally, not perceptually. They do not see or hear the same thing; they merely react to one another's anxiety. Well-known examples include the 1962 Tanganyika laughter epidemic, in which hundreds of schoolgirls in Tanzania began laughing uncontrollably after one girl initiated the episode.[85] Another example occurred in 1983 in the West Bank, where a rumor of a toxic gas attack led to fainting and nausea among hundreds of students, even though no poison was ever present.[86] In both cases, the symptoms spread through social influence, but there were no shared perceptions or visions.

Therefore, the claim that multiple people could all see the same supernatural event through "mass hallucination" is not only unsupported — it is scientifically incoherent. Hallucinations are private mental events occurring within individual brains. For two or more people to experience the same hallucination simultaneously, their neurological activity would have to synchronize in detail down to the sensory cortex — an impossibility without an external cause. If dozens or hundreds of people report witnessing the same miracle, the only plausible explanations are that they jointly perceived a real external event or that they conspired to fabricate it. The latter, in the case of independent witnesses across different backgrounds and interests, quickly becomes less plausible than the miracle itself.

It is also worth noting that historians routinely accept mass testimony for secular events — battles, natural disasters, coronations — without invoking hallucination or hysteria. Only when the event involves a miracle do skeptics reach for these unscientific explanations. This inconsistency exposes the real issue: not an evidentiary problem, but an ideological one. When multiple people report something that fits the secular worldview, it is "collective memory." When they report something that does not, it becomes "mass delusion." In reality, there is not a single documented case in human history where an entire group of sane individuals has shared an identical, detailed hallucination. The supposed "mass hallucination" theory therefore explains nothing — it simply denies what is uncomfortable. The irony is that this denial, not belief in miracles, represents the true leap of faith.

Eyewitness Discrepancies and the Myth of Infallible History

A frequent tactic among skeptics attempting to dismiss reports of miracles is to highlight contradictions or minor discrepancies among eyewitness accounts. They argue that because witnesses sometimes report events differently, the events themselves must be false or fabricated. While this may appear compelling, it overlooks a crucial reality: such inconsistencies are a universal feature of all historical testimony, whether the events are miraculous or entirely natural.

Even for ordinary occurrences—such as battles, political decisions, or natural disasters—eyewitnesses rarely produce perfectly harmonized accounts. Differences in perspective, timing, memory, and interpretation naturally lead to variations in reporting. Yet, historians do not conclude that these events did not happen simply because witnesses differ on details; instead, they cross-examine sources, assess plausibility, and seek corroborating evidence to reconstruct events as accurately as possible.[87] The same methodological principles apply equally to accounts of miracles.[88]

For example, consider accounts of a historical event witnessed by multiple observers. Individuals might describe it differently, emphasizing different aspects or recalling sequences in varying orders. These divergences do not cast doubt on the reality of the event itself; they merely reflect the normal human limitations of perception and recollection. The same is true for miracles. Minor differences among witnesses—whether in the sequence of events, the number of observers, or the exact words spoken—cannot logically serve as grounds for outright denial.

Skeptics often treat miracle reports as if they were required to meet the impossible standard of a laboratory experiment, where every detail is perfectly recorded and repeatable. History, however, operates in the real world of human experience, where variations in testimony are inevitable. To demand infallibility for miracles while excusing the same imperfections in natural events reflects a double standard rather than an objective assessment. In other words, contradictions in eyewitness accounts do not disprove miracles; they are simply part of the normal fabric of historical reporting.

Conclusion

In summary, the common objections to miracles—ranging from appeals to natural law, alleged impossibility, or claims of mass delusion—fail under careful historical scrutiny. Skeptical arguments often rely on double standards, demanding infallibility from miraculous reports while excusing the same imperfections in ordinary events. Contradictions in eyewitness testimony, variations in detail, and the passage of time do not invalidate historical evidence; they are normal aspects of human reporting. Just as historians reconstruct natural events from imperfect sources, the accounts of miracles can be assessed with the same critical tools. Having addressed these general contentions, we now turn to specific Islamic miracles and select non-Islamic examples, examining the historical evidence and arguments for their authenticity. This will allow a detailed evaluation grounded in actual reports rather than abstract objections.

Evolution: The Modern Myth — Philosophy, Evidence, and Blind Faith

Few ideas in modern history have shaped the Western mind as deeply as evolution. Presented as settled science, it functions not merely as a biological theory but as the foundational belief of the secular worldview—the story of how life came to be without design, purpose, or Creator. From classrooms to popular media, evolution is treated as a self-evident truth, and dissent is often dismissed as ignorance or fanaticism. Yet beneath its cultural prestige lies a complex mixture of philosophy, assumption, and selective reasoning.

This chapter examines evolution from both scientific and philosophical angles. It begins by exploring its widespread acceptance and ideological role, then distinguishes between what evolution actually claims and the evidence offered for it. From there, we will analyze its mechanisms, logical foundations, and evidentiary weaknesses, as well as the assumptions and deceptions that sustain it. The goal is not to attack science but to question whether evolution truly meets its own standards of evidence, coherence, and intellectual honesty.

Evolution as Orthodoxy: The Creed of the Modern West

Few ideas have permeated Western thought as deeply as evolution. From the earliest stages of schooling, it is taught not as one theory among others but as the unquestionable framework for understanding life. Textbooks, museums, and media all present it as fact, shaping how generations perceive nature, humanity, and purpose. To doubt it is often seen as rejecting reason itself.

By the time students finish secondary school, they have absorbed a seamless narrative—from the Big Bang to modern humans—with no serious mention of alternative explanations. “Education” in the modern sense has come to mean acceptance of evolution.

This dominance extends far beyond academia. Popular science, documentaries, and entertainment all reinforce the evolutionary story as self-evident. Repetition and institutional authority have made evolution more than a scientific theory—it is a cultural constant, a belief so embedded that most people accept it unconsciously.

Its power lies in this invisibility. Most in the modern West do not choose to believe in evolution; they inherit it as part of their intellectual environment. It functions as a silent premise in discussions of biology, ethics, and even politics, defining what counts as “educated” thought while casting dissent as irrational or unscientific.

In the modern West, belief in evolution has become a symbol of intellect and progress. Accepting it is seen as a sign of rationality, while rejecting it is treated as regression into ignorance or superstition. This perception is not the result of scientific argument but of social signaling—to affirm evolution is to align oneself with modern science, technology, and the prestige of the academic establishment.

This association took root during the late nineteenth and early twentieth centuries, when Darwin’s ideas were championed as proof that humanity had finally escaped religion. Evolution became synonymous with the scientific spirit itself—the mark of an enlightened mind that no longer needed divine authority. To question it, therefore, was to question the very narrative of human progress.

As a result, public debates about evolution rarely concern evidence. They concern identity. Belief in evolution is often less about biology and more about belonging—about being on the “right side” of history, modernity, and reason. Within this framework, even mild skepticism is not treated as a scientific position but as a moral failing, a betrayal of the modern worldview that defines itself in contrast to faith and tradition.

For much of Western history, the Biblical creation account provided the framework for understanding human origins and purpose. With the rise of Darwinism, that role was quietly replaced. Evolution became the genesis story of the modern secular world—a narrative that explains life, order, and complexity without reference to God.

In this new story, humanity is not the deliberate creation of a divine will but the accidental product of natural processes. This shift transformed more than biology; it redefined the moral and metaphysical foundations of Western thought. Evolution offered what religion once did: an origin myth that explained who we are, where we came from, and why we exist. The difference was that this story claimed scientific authority instead of revelation.

Because of this, evolution occupies a place in secular culture that goes beyond science. It satisfies the same emotional and philosophical needs that creation once did, offering meaning, continuity, and even a sense of destiny. The reverence with which it is treated—the resistance to any challenge or alternative—reflects not only intellectual confidence but also existential dependence. If evolution were proven false, the secular worldview would lose its creation story, and with it, its claim to self-sufficiency.

The near-religious reverence for evolution in the modern West is sustained not only by emotion but by powerful institutions. Once evolution replaced divine creation as the accepted origin story, it gained a sacred status within secular culture. To affirm it became a moral and intellectual duty; to doubt it, a form of heresy.

This reverence is maintained through an interlocking network of schools, universities, and media that present evolution as settled truth.

Educational systems rarely encourage genuine debate; they transmit consensus. Museums display evolution as visual fact, and documentaries narrate it with the same solemn tone once reserved for scripture. The constant repetition gives the impression that belief in evolution is the only rational position available.

Most people accept it not through personal investigation but through trust in authority. Few have examined the data or studied the mechanisms; they simply defer to "the experts." In practice, evolution's credibility rests less on evidence accessible to the public than on the prestige of scientific institutions. This dynamic mirrors religious faith in its structure—belief based on institutional authority, emotional assurance, and the comfort of social conformity.

Thus, evolution's power does not lie in scientific argument but in the machinery that enforces its credibility. It is believed widely because it is taught universally, and it is taught universally because it is believed widely—a self-reinforcing cycle that keeps dissent outside the bounds of acceptable thought.

In the modern West, questioning evolution carries social and professional consequences that reveal its cultural sanctity. Those who express skepticism—whether scientists, teachers, or students—quickly find themselves marginalized. Careers have been stalled, research grants withdrawn, and reputations destroyed for even modest criticism of Darwinian assumptions.[89] The response to dissent often has little to do with science and everything to do with enforcing ideological conformity.

The stigma operates through ridicule and exclusion. To question evolution is to invite the label of "anti-science" or "religious fundamentalist," regardless of one's reasoning or credentials. This social pressure effectively silences debate before it begins. In academic settings, the expectation to conform is so strong that many who harbor doubts remain silent, understanding that professional survival depends on submission to the prevailing orthodoxy.

Belief in evolution has thus become more than a scientific position—it is a cultural identity marker. Affirming it signals allegiance to secular modernity, rationalism, and progress. Rejecting it, conversely, aligns one with religion, "backwardness," and superstition. The divide is emotional and moral before it is intellectual. In this way, evolution functions less as an empirical theory and more as a tribal boundary, separating those who belong to the modern secular order from those who do not.

A striking feature of the cultural treatment of evolution is the selective application of skepticism. In nearly every other domain of science, questioning, debate, and critical scrutiny are encouraged—even expected. Yet evolution is largely shielded from the same rigorous doubt, creating a double standard in scientific reasoning.

For example, medical studies, climate research, and technological claims routinely undergo intense peer review and public scrutiny. Errors, biases, and alternative interpretations are openly discussed. By contrast, evolution is often presented as settled truth, with dissenters immediately marginalized or dismissed. This discrepancy reveals that acceptance of evolution is not purely a matter of evidence, but also of cultural and ideological enforcement.

The contrast highlights how belief in evolution functions less as a product of empirical reasoning and more as a marker of conformity to a dominant worldview. The very principles of critical inquiry—questioning assumptions, evaluating data, testing alternative hypotheses—are often suspended when it comes to evolution, underscoring its status as a modern orthodoxy.

The cumulative effect of cultural dominance, institutional reinforcement, social pressure, and selective skepticism is that evolution functions more like a creed than a scientific theory. It is treated with reverence, defended against dissent, and woven into the identity of those who accept it. Its authority is enforced socially and institutionally, and questioning it is often equated with immorality, ignorance, or backwardness.

Like religious dogma, evolution maintains its power through a combination of repetition, institutional support, and emotional commitment. The theory is rarely challenged in public discourse, and its foundational assumptions—such as the inevitability of naturalistic explanations—are seldom scrutinized. In this sense, evolution has transcended its original scientific purpose, becoming a cultural orthodoxy: a lens through which nature is interpreted and through which intellectual belonging is determined.

The parallels are clear: belief is expected, dissent is punished, and adherence is socially rewarded. By framing evolution as unquestionable truth, the modern West has created a system in which scientific claims, philosophical commitments, and cultural identity are inseparably linked—a system that mirrors the structures and psychological dynamics of traditional religious faith.

Mechanism and History: Clarifying Evolution

To understand evolution properly, it is essential to distinguish between mechanism and history. The mechanism refers to the processes through which life changes over time: mutations, natural selection, genetic drift, and other forces proposed to drive the transformation of organisms. It asks the question: how could life change from one form to another? History, by contrast, refers to the actual sequence of events—the lineages, ancestral relationships, and order in which species appeared. It asks what changes actually occurred and which organisms are related to which. While mechanism concerns possibility, history concerns factual reconstruction of life's past.

This distinction is critical because many discussions of evolution conflate the two, treating evidence for one as automatic proof of the other. Without it, critiques of historical reconstructions are often misunderstood as critiques of the mechanisms themselves, and vice versa. A clear separation allows for a more precise and logical evaluation of evolutionary claims and prevents common misunderstandings.

In practice, most popular arguments for evolution focus on the history of life rather than the plausibility of its mechanisms. Textbooks, documentaries, and media presentations emphasize universal common ancestry, the tree of life, and the order of species, creating the impression that these reconstructions confirm how life changes. This focus often obscures the more fundamental question: can natural mechanisms plausibly produce the diversity of life? Conflating mechanisms and history has led to widespread misconceptions. Historical reconstructions are frequently treated as direct proof that the proposed mechanisms are correct, while the plausibility of the mechanisms is often assumed based on acceptance of the historical narrative.

Importantly, one can accept certain elements of evolutionary theory without accepting others. For example, the fact that universal common descent does not automatically imply that life evolved through unguided physical processes has been recognized by some eminent evolutionary biologists. Here is a reference regarding this fact:

"Ernst Mayr, one of the 20th century's leading evolutionary biologists, divides Darwinian evolutionary theory into five distinct parts. First (1), there is the general thesis of evolution as such. This states the changeability of species as opposed to a constant, unchanging world.

Second (2), there is common descent, the thesis that all different branches of animal species share a common ancestor. Third (3), there is the thesis of gradualness, which states that species change by small, successive and cumulative mutations. Thus there are no big 'jumps' in evolution where a completely new form of animal is formed through just one mutation. Fourth (4), there is the thesis of populational speciation, which states that new species can emerge from existing populations through changes. Fifth (5), there is natural selection working on random mutations – the Darwinian mechanism for explaining evolutionary adaptations. As Mayr argues, it is in principle possible to accept just parts of Darwinian evolutionary theory. For example, many of Darwin's contemporaries did not accept natural selection as the mechanism of evolutionary change, but accepted the thesis of common descent."[90]

This distinction underscores the importance of separating mechanism from history. Acceptance of historical claims, such as common descent, does not require acceptance of any particular mechanism. Likewise, one could explore mechanisms of change without committing to a particular reconstruction of life's history. Understanding this allows readers to critically assess the arguments for Evolution without falling into the common trap of conflating two distinct questions.

By keeping mechanism and history distinct, discussion of evolution becomes more precise. The distinction allows readers to evaluate whether life could change naturally and what actually happened in the past. This conceptual clarity lays the foundation for a rigorous and fair examination of Evolution.

The Burden of Proof

It is often taken for granted that the burden of proof lies with anyone who challenges evolution, a presumption that seems to extend to virtually anything endorsed by Western academia. This assumption is widespread, yet it rests on very shaky grounds. The burden of proof is not determined by the popularity of a belief, the authority of the institution promoting it, or the number of people who accept it; it is a principle of rational inquiry that is assessed independently, before any evidence is considered, not after. To claim that one side automatically carries the burden simply because it opposes a mainstream view is not only logically flawed but also intellectually lazy, as it bypasses critical evaluation and substitutes authority for reason.

Compounding this is a related, but equally mistaken, assumption: that any explanation which challenges religious belief, or appears to contradict God, carries no burden of proof, while the onus automatically falls on those who defend materialist or scientific explanations. This notion is deeply flawed and intellectually untenable. It effectively grants automatic credibility to claims solely because they touch on the divine or the supernatural, regardless of whether they are supported by evidence, logical reasoning, or consistent observation. Such a framework undermines rational discourse entirely, replacing critical evaluation with arbitrary privilege. Any honest and rational thinker must reject these assumptions outright, recognizing that the burden of proof applies universally without bias towards one ideology or belief system.

A common justification offered for placing the burden of proof on those who challenge evolution is rooted in the observation that unguided physical processes, such as natural selection and genetic variation, do explain at least some aspects of life. From this starting point, it is argued that the simplest explanation is to assume these processes account for all aspects of biological complexity, and therefore anyone invoking additional mechanisms, purposes, or entities bears the responsibility of providing evidence for them. This line of reasoning appeals to a principle often described as Occam's Razor, where simplicity is taken as a guide to plausibility.

However, this reasoning is not as decisive as it is often presented. While it is true that natural processes demonstrably account for many phenomena, the fact that they explain part of the picture does not automatically justify the conclusion that they must explain everything. Assuming universal sufficiency from partial success is a logical leap, not a proof. Just because a hammer can drive some nails does not mean it is the only tool that can ever be needed for construction. In the same way, demonstrating that evolution explains certain features of life does not automatically negate the possibility that additional factors—whether known or yet to be discovered—could play a role. The burden of proof does not shift by default; it rests on the claimant asserting something beyond what is already established.

This discussion regarding physical processes is deeply problematic for another reason: it implicitly assumes that the operation of physical processes contradicts the existence of God or the possibility that life was designed. Such an assumption is far from self-evident, and it introduces a bias that colors the entire argument about the burden of proof. It presumes that accepting natural mechanisms as part of life's development automatically excludes purpose or design, which is a significant philosophical claim in itself. The implications of this assumption are complex, and this topic will be explored in depth in the next section of this chapter, where we examine the relationship between physical processes, design, and the broader questions about life's origin.

One of the biggest problems with the usual argument about the burden of proof comes when we look closely at the similarities we see between living things. Evolutionists often argue that so many species share traits that it couldn't just happen by chance. They say the simplest explanation is common descent.[91] At first, that sounds reasonable, and it's repeated constantly in textbooks and popular science articles as if it's a slam-dunk proof of evolution. But when you dig deeper, it's not so simple.

The problem is that there are many similarities in living things that even evolutionists admit cannot be explained by common descent alone.[92] To deal with these, they invent extra explanations, such as convergent evolution, functional constraints, or adaptive pressures.[93] Each of these is an extra layer added on top of the basic evolutionary story, and none of them are as solidly evidenced as the original idea of natural selection. In other words, the theory has to keep patching itself with additional assumptions to make the observed patterns fit. On the other hand, the idea of common design can account for both kinds of similarities—the ones evolutionists can explain with common descent and the ones they cannot—without needing to invent multiple fixes. [94] If you judge by who has to explain more, the burden of proof should logically fall on the Evolutionists based on the criteria that they themselves were claiming to hold.

Another common move is the parsimony argument. Evolutionists often claim that design is automatically less credible because it introduces an extra entity—the designer—while evolution supposedly relies only on natural processes.[95] On the surface, this looks neat. It's like saying, "Evolution is simpler, so it's better." But it doesn't hold up under scrutiny. If evolution is supposed to be true, it implies a massive number of additional species that we don't see in the fossil record[6]. To make the story work, evolutionists have to assume these species existed anyway, even though we have no direct evidence of them. In practice, that means evolution is introducing far more "extra entities" than a single designer would. So by the same logic they use against design, the burden of proof should fall squarely on evolutionists.

The consequences of this go far beyond academic debate—they reveal a clear double standard in how evolutionists treat evidence and assign the burden of proof. Critics of evolution are expected to prove something almost impossible: that evolution cannot explain everything, or that design is a plausible alternative. Meanwhile, evolutionists themselves are allowed to invent extra explanations whenever the evidence doesn't fit, whether it's convergent evolution, adaptive constraints, or missing fossil species. These ad hoc fixes are added freely, without the same level of scrutiny, and yet no one demands that evolutionists provide independent evidence for every single hypothetical species or mechanism they invoke. This is a textbook example of hypocrisy: one side is held to impossible standards, while the other side gets a free pass to adjust the theory whenever it runs into problems.

This double standard becomes especially obvious when we examine the argument about similarities between species. Evolutionists claim that shared traits prove common descent, yet they simultaneously admit that many similarities cannot be explained this way. Instead of accepting that this is a problem for the theory, they create additional explanations and expect critics to ignore the inconsistencies. At the same time, design proponents are criticized for "introducing an extra entity"—the designer—even though evolutionists themselves rely on countless unobserved species and unseen processes to make the theory appear consistent. By any fair measure, the burden of proof is applied inconsistently. Recognizing this hypocrisy is crucial: it shows that the debate over evolution is not just about evidence, but also about how the rules of reasoning and proof are selectively enforced.

Another point that exposes this inconsistency is the way the burden of proof is treated when it comes to miracles. Whenever someone argues that a miracle occurred, the immediate response from the same scientific establishment is that the burden of proof lies on the person claiming it. The reasoning is always the same: "We don't see miracles happening now, so extraordinary claims require extraordinary evidence." In other words, the claim is treated as so unusual and outside ordinary experience that it must be proven beyond all doubt before it can even be considered.

But notice how this exact same reasoning, if applied consistently, would put the burden of proof squarely on the person arguing for evolution. No one sees evolution happening now in the way the theory describes it—entirely new body plans emerging, one species turning into another, or completely new genetic information arising spontaneously. What we see today are small variations within limits, but not the large-scale transformations the theory depends on. The great changes evolution claims to have produced supposedly happened in the distant past, under conditions no one has ever observed. So, by the same principle used against miracles—that something not observed now carries the burden of proof—it is evolutionists who should have to prove their claims, not those who question them.

The irony is hard to miss. When it comes to miracles, the scientific community insists that unobserved events must be treated with maximum skepticism. Yet when it comes to evolution, which depends entirely on unobserved and unrepeatable events, the same people declare it the default truth. In both cases, the reasoning should be identical: if something cannot be observed, repeated, or directly demonstrated under the conditions it supposedly occurs, the burden of proof lies on the person making the claim. But this rule is applied selectively—strictly enforced against anything that supports belief in God, and conveniently ignored when it supports materialism.

This selective skepticism once again reveals that the burden of proof in debates about evolution is not being assigned based on logic or fairness but on philosophical bias. It is not about what the evidence shows, but about which worldview someone wants to protect.

Physical Processes

One of the most important points in this debate is the relationship between physical processes and evolution. A very common assumption among evolutionists is that examples such as bacteria developing antibiotic resistance prove that life emerged purely through unguided physical processes, without any design or purpose involved. This assumption is repeated so frequently that many take it as self-evident—but it is deeply mistaken.

The underlying belief is that if God were creating life, each species (or at least major groups of species) would have to be formed through special, direct creation, with no physical processes involved in shaping their features. In other words, evolutionists imagine that design and natural processes are mutually exclusive explanations. This is a false dichotomy—and, frankly, one of the weakest assumptions made by proponents of evolution. It is closely tied to what is often called the "God of the gaps" argument: the claim that intelligent design is invoked only where physical explanations fail. In reality, this framing is entirely backwards.

The error becomes obvious with a simple analogy. Suppose I bake a cake using an oven. I gather ingredients through certain physical processes, mix them using mechanical actions, and the heat of the oven—another physical process—causes the mixture to transform into a baked cake. There is no question that physical processes are involved. But does their presence mean there is no intelligent agency behind the cake's creation? Of course not. The fact that natural mechanisms are used does not eliminate the role of design; it simply shows how the design is implemented. In this scenario, I am the intelligent cause directing those processes toward a purposeful outcome. The same reasoning applies to life: the operation of physical laws and mechanisms does not in any way negate the involvement of an intelligent Creator.

A further mistake made by evolutionists is the unwarranted leap from "physical processes occur" to "these processes are blind and unguided." Observing that something happens through natural mechanisms does not prove that those mechanisms lack purpose or direction. There is no empirical evidence that the processes driving biological adaptation—whether in bacteria, plants, or animals—are genuinely random or purposeless. That assumption is philosophical, not scientific.

Consider another analogy. Imagine an automated factory run by artificial intelligence. Once set up, the factory produces items through entirely physical and mechanical processes. The machines operate consistently, and the assembly lines move efficiently without any direct human intervention. But does that mean no intelligence was behind the factory or its operations? Clearly not. The entire system, including the algorithms that regulate it, reflects intentional design. The fact that it now functions autonomously through physical means does not erase the intelligent cause that brought it into existence.

In the same way, the presence of regular, natural processes in biology does not imply the absence of a designer. Rather, it may testify to the sophistication and foresight of that design. The idea that physical causation and intelligent purpose are incompatible is not a conclusion drawn from evidence—it is an assumption born from a materialist worldview. Once that assumption is exposed, the supposed conflict between design and physical processes dissolves entirely.

The Illusion of Infinite Change

One of the most common justifications offered for evolution is the observation that small, gradual changes occur in living organisms today.

It is argued that if such changes continue over vast periods of time, they could accumulate to produce all the diversity of life we see. Here is a reference for this fact:

"Evolutionary biologist Mark Ridley summarizes four main lines of evidence for Darwinian evolutionary theory. First (1), small-scale changes in organisms can be directly observed. It is then extrapolated that past changes in organisms can be explained through the same evolutionary mechanisms observed in the present."[96]

This argument appears persuasive at first glance, because it appeals to something observable—variation within species—and then extends it to explain everything from bacteria to human beings. However, this reasoning rests on a critical assumption: that the limited changes we observe are not only possible but capable, in principle, of producing unlimited transformation. That assumption is never demonstrated; it is simply asserted.

This reveals a striking inconsistency in standards of proof. When theists speak of miracles or divine intervention, critics insist that such events must first be shown to be possible before any evidence can even be considered. Yet when evolutionists propose that small variations can accumulate without limit, no such demand is made. The possibility of limitless transformation is taken for granted, despite the absence of any empirical or logical justification. The same people who require proof of possibility from believers are unwilling to meet that same standard for their own claims.

Even more remarkable is the assumption that the mechanisms driving these small changes have no inherent boundaries. Everywhere in nature we see limits: physical limits, chemical limits, genetic limits, developmental limits. Every natural system operates within certain constraints—materials can only stretch so far, organisms can only survive within certain environmental conditions, and biological processes function within narrow biochemical tolerances. Yet for this one process—evolution—it is assumed that there are none. We are told that a process responsible for minor adaptations, like a change in beak shape or color variation, can also redesign entire body structures, invent new organs, and rewire complex biological systems, all without guidance or planning.

This supposed exception to every other law of limitation in nature is treated as if it were a miracle—but a miracle of materialism, not of God. The irony is profound: those who mock theists for believing in miracles insist that evolution alone performs one of its own, creating complexity, intelligence, and design from randomness and blind forces. The double standard could not be more obvious.

In reality, the mechanisms observed in minor adaptation are fundamentally incapable of driving large-scale evolutionary change. The small adjustments we see—often caused by the modification or loss of existing genetic information—do not explain how entirely new information arises. Major changes in body structure would require mutations that act early in embryonic development, altering the body plan itself.

Here is the problem though. These early-acting mutations are always damaging to the organism and hence are not beneficial mutations that can be selected by natural selection. As evolutionary biologist R.A. Fisher explains such mutations are "either definitely pathological (most often lethal) in their effects," or they result in an organism that cannot survive "in the wild state."[97] This problem has been called a "great Darwinian paradox"[98] by geneticist John F. Mcdonald who explains the problem:

"Those [genetic] loci that are obviously variable within natural populations do not seem to lie at the basis of many major adaptive changes, while those loci that seemingly do constitute the foundation of many if not most major adaptive changes are not variable within natural populations."[99]

This evidence shows that the mechanisms capable of producing small, survivable variations are not the same as those required for large-scale transformations. The mutations that could theoretically produce new body plans or major innovations are precisely the ones that destroy viability. This is not a technical obstacle—it is a structural limit inherent to biological systems.

What we observe, therefore, is not an open-ended process capable of infinite creativity, but a system designed for limited variation within established boundaries. The claim that these modest variations can accumulate into entirely new forms of life is not an empirical conclusion but an assumption—an act of faith disguised as science. The illusion of infinite change remains just that: an illusion, sustained by philosophical commitment rather than observable evidence.

The Myth of Universal Common Ancestry

When examining the scientific arguments most frequently presented in favor of evolution, one quickly realizes that the vast majority of them do not actually concern the mechanism of evolution, but rather the idea of universal common ancestry — the claim that all living things ultimately descend from a single organism. Nearly every line of "evidence" offered by evolutionary literature or popular science sources centers around some kind of similarity among living beings. These include genetic similarities, morphological patterns, embryological resemblances, and biogeographical distributions, all of which are interpreted as proof that life must share a common origin. Here are two references regarding this fact:

"If all these species have a common ancestor, the lineages descending from that common ancestor had to start with the same trait value. It follows that the task of explaining why vertebrates have camera eyes is essentially connected to the task of explaining why other groups have other kinds of eyes while still others have none at all. Given how central the thesis of common ancestry is to evolutionary reasoning, one might expect there to be a vast literature in which the evidence for that claim is amassed. In fact, the question is discussed, but the literature on it is hardly vast. For most evolutionists, the similarities that different species share make it obvious that they have common ancestors, and there is no reason to puzzle further over the question. The kind of genealogical question that attracts far more attention in evolutionary biology concerns how various species are related to each other, not whether they are."[100]

"Similarity, ergo common ancestry. This form of argument occurs so often in Darwin's writings that it deserves to be called modus Darwin. The finches in the Galapagos Islands are similar; hence, they descended from a common ancestor. Human beings and monkeys are similar; hence, they descended from a common ancestor. The examples are plentiful, not just in Darwin's thought, but in evolutionary reasoning down to the present (Sober 1999a)."[101]

Yet upon closer inspection, these are not a multitude of independent evidence. They are, in fact, variations of a single argument: that similarity implies common descent. No matter how it is framed — whether through DNA sequences, homologous structures, or developmental stages — it is the same basic reasoning repeated in different forms. The supposed "mountain of evidence" for evolution, therefore, largely collapses into a single philosophical claim dressed up in scientific terminology.

This pattern is easy to notice when observing how leading advocates of evolution or atheism defend their position. When confronted with arguments for design or divine creation, they often revert to examples of similarity: "If God created life, why would He use the same DNA code in all creatures?" or "Why do humans share over 90% of their genes with chimpanzees?" The implicit assumption is that a Creator would make all species completely distinct, and that any resemblance must therefore prove descent. In reality, these are not multiple lines of evidence but merely repetitions of the same idea — that likeness equals lineage — an assumption rather than an established fact.

The first thing I want to discuss is the claim that these similarities in living things cannot be due to chance. Virtually no one — not even the evolutionists themselves — accepts this as a rational explanation, as can be seen from the references cited earlier. However, the reasoning behind their rejection of chance reveals deep inconsistencies within evolutionary epistemology. Their stance against chance is not grounded in consistent logic but rather in selective application of probability, depending on what suits their argument at a given moment.

The first reason for this is related to the fine-tuning of the universe. This is a very important concept in Physics and is routinely discussed in debates regarding the existence of God. I will just give one reference regarding this topic:

"We live not only on a friendly planet but also in a friendly universe. It follows from the fact of our existence that the laws of physics must be friendly enough to allow life to arise. It is no accident that when we look at the night sky we see stars, for stars are a necessary prerequisite for the existence of most of the chemical elements, and without chemistry there could be no life. Physicists have calculated that, if the laws and constants of physics had been even slightly different, the universe would have developed in such a way that life would have been impossible. Different physicists put it in different ways, but the conclusion is always much the same."[102]

If the fundamental constants or physical laws of our universe were even slightly different, the existence of complex life would be impossible. The sheer improbability of these precise values — perfectly tuned for life — has long been cited as compelling evidence of intentional design rather than chance.

Yet many atheists and evolutionists attempt to explain away this fine-tuning through the multiverse hypothesis. This idea proposes that there are an infinite or near-infinite number of universes. If so, then improbable conditions such as ours would eventually occur somewhere purely by chance. Within such a framework, anything can be explained probabilistically, since in an infinite multiverse, even the most unlikely events become inevitable.[103]

Here lies the contradiction: many evolutionists who accept the multiverse as a rational explanation for cosmic fine-tuning simultaneously argue that the similarities among living things cannot be due to chance. But in an infinite multiverse, nothing is truly improbable — not even the spontaneous emergence of life or identical biological patterns appearing independently. If chance can explain fine-tuning, then by the same logic, it can explain any degree of biological similarity. Thus, by their own reasoning, all probability-based arguments for evolution lose their force in an infinite multiverse.

This epistemological inconsistency is not limited to those who accept the multiverse. Even evolutionists who reject it often appeal to deep time — the notion that "given billions of years, anything can happen." This argument is invoked repeatedly to defend the possibility of life arising through unguided processes. When confronted with the absence of observed large-scale evolutionary change, the answer is the same: "billions of years." It is claimed that, although we cannot see major evolutionary transitions occurring today, vast stretches of time make such transformations inevitable. But if the passage of immense time can render this unlikely event possible, then why should the appearance of similarities through chance be excluded from that same logic? Either deep time allows improbable events to happen — or it does not. Evolutionists cannot rationally invoke it when it helps them and dismiss it when it does not.

The problem becomes even more evident when considering convergent evolution — the frequent occurrence of strikingly similar features arising independently in unrelated organisms.[104] Evolutionists themselves admit that similar biological structures, functions, and even genetic sequences often appear in species that supposedly share no recent common ancestor. Examples abound: the camera-like eyes of octopuses and vertebrates, the streamlined bodies of dolphins and ichthyosaurs, or the genetic and behavioral parallels found in marsupial and placental mammals. According to evolutionary theory, these similarities evolved multiple times, independently, through purely natural processes.

In other words, evolutionists are forced to concede that similarities can and do appear without common descent, and not rarely, but routinely. Yet when it comes to evidence for universal common ancestry, they turn around and claim that such similarities are so unlikely to arise by chance that they must be due to descent from a shared ancestor. This is another blatant contradiction. If random mutations and natural selection can supposedly produce the same complex traits again and again in unrelated lineages, then how can anyone argue that similarities across life forms are too improbable to have arisen without common descent?

In summary, I do not accept chance as a valid explanation for the similarities in living organisms, but my rejection is based on a consistent epistemology — one that recognizes limits to what unguided processes can reasonably accomplish. The evolutionists' rejection, however, is based on inconsistency and convenience. When chance supports their theory, it becomes a legitimate explanatory tool; when it undermines it, chance is suddenly unscientific. This double standard reflects not strength but weakness — an unstable foundation built on shifting arguments rather than consistent logic.

Many evolutionists frame the argument for common ancestry around improbability: the similarities among living things are so unlikely to occur by chance that they must have a common origin. Even if one were to accept this claim, it does not logically follow that these similarities are therefore due to common descent. Improbable similarities can arise in only two ways: either by common descent or by common design. The improbability of chance alone does not validate evolution's preferred explanation. With this in mind, several reasons show that common design is a plausible — and indeed better — explanation for these similarities.

Evolutionists often dismiss common design as an ad hoc explanation, claiming that similarities would be unexpected if intelligent design were true. Yet this is contrary to everyday experience. Whenever a single designer creates multiple products, similarities frequently appear. A software developer producing multiple programs leaves a signature style in code; a writer produces texts with recurring linguistic patterns; an artist creates works with recognizable motifs; an architect designs buildings with shared structural or aesthetic features. By analogy, it is entirely plausible that similarities among living organisms reflect the actions of a common designer rather than descent from a single ancestor.

Evolutionists often argue that common descent is obvious because of observed patterns of similarity. Yet these patterns — nested hierarchies and degrees of likeness among species — were recognized long before Darwin, and interpreted within a framework of divine creation. Linnaeus, for example, systematically classified organisms into genus and species, observing clear hierarchical patterns across plants and animals, which he understood as reflecting a divine plan. This shows that the intuition that similarity implies descent is not universal, but historically and culturally conditioned. The supposed self-evidence of common descent is therefore far weaker than evolutionists claim.

One of the most glaring problems for common descent is the phenomenon of convergent evolution: strikingly similar traits appear independently in unrelated organisms. Examples include the camera-like eyes of octopuses and vertebrates, the streamlined bodies of dolphins and ichthyosaurs, or the repeated evolution of wings in bats, birds, and insects. Evolutionists are forced to invoke multiple independent events to explain these patterns, whereas common design explains all similarities — those attributed to descent and those arising independently — under a single principle: a single designer created similar solutions to similar functional challenges.

Evolutionists often appeal to a version of Occam's Razor in their own arguments: the simplest explanation that accounts for the evidence is preferred. Yet when common descent requires multiple explanations — descent plus convergent evolution — to account for all similarities, and common design explains everything in a single step, it is clear which explanation is simpler. By their own standard, common design is the more parsimonious and coherent explanation.

Because common design explains all observed similarities with a single, consistent principle, it directly challenges the evolutionist claim that universal common descent is overwhelmingly supported by evidence.

The improbability of chance alone does not establish descent; and once one allows for common design, the central argument for universal ancestry collapses.

In conclusion, while skeptics and evolutionists often complain about philosophy, logic, or reasoning when discussing miracles or divine action, their own reasoning about life is a stark contrast: the logic of evolution is inconsistent, selective, and full of double standards. When applied rigorously, the very arguments evolutionists use to defend common descent reveal their hypocrisy and intellectual dishonesty, exposing the weakness of the case for universal common ancestry.

The Fossil Record: Evidence and Interpretation

Fossils are the preserved remains or traces of organisms from the past, typically found in sedimentary rock. The fossil record is the accumulation of all discovered fossils, providing a chronological snapshot of life on Earth across millions of years. Scientists use the fossil record to study how life has changed over time and to reconstruct the history of organisms and ecosystems. When evolutionists present the fossil record as evidence for evolution, they typically make two types of arguments.

This argument focuses on transitional fossils — specimens that appear to show intermediate characteristics between different species. These are often presented as evidence that one species gradually evolved into another. However, this is essentially the same argument from similarity that we discussed in the previous section on universal common ancestry. Claiming that a fossil is "transitional" because it shares features with multiple species is just a repetition of the earlier argument that similarities imply evolutionary descent. As we have already seen, such arguments are not independent evidence for evolution but merely variations on the same reasoning, which faces serious philosophical and empirical problems.

•

The second type of argument from the fossil record is related to the mechanisms of evolution. This includes claims that the fossil record demonstrates a gradual pattern of change over time, consistent with the slow accumulation of evolutionary modifications. These arguments aim to connect fossils not just to ancestry, but to the processes — natural selection, mutation, and other mechanisms — by which evolution supposedly occurs. This is the type of evidence we will address in this section, as it pertains specifically to the mechanisms of evolution rather than merely the pattern of similarities among organisms.

Evolutionists argue that the fossil record provides evidence for the mechanism of evolution through a temporal sequence of fossils. According to this view, fossils appear in chronological order with slight variations between successive forms, creating the appearance of gradual change from one species to another. Here is a reference the argument from an Evolutionist:

"What Darwin didn't have were enough fossils to show clear evidence of gradual changes within species, or of common ancestors. But since his time, paleontologists have turned up fossils galore, fulfilling all the predictions mentioned above. We can now show continuous changes within lineages of animals; we have lots of evidence for common ancestors and transitional forms (those missing ancestors of whales, for instance, have turned up); and we have dug deep enough to see the very beginnings of complex life."[105]

Evolutionists claim that these small variations can arise through random mutations, while natural selection preserves those mutations that are beneficial, eventually producing entirely new forms of life. This process — often called the primary mechanism of evolution — is central to the modern evolutionary framework, although some biologists have proposed alternative mechanisms as primary contributors to evolutionary change.

The core of the argument is straightforward. When fossils of similar species appear sequentially in the geological record, evolutionists see a pattern suggesting transformation over time. The claim is that these sequential fossils, closely related in appearance and arranged in order through time, look like evolution is occurring, and that this visual pattern alone is compelling evidence.

The logic behind this argument relies on a particular application of Occam's Razor: if something looks like a certain process is occurring, then the simplest explanation is that the process is indeed taking place. In the case of fossils, evolutionists argue that the sequential, gradual changes visible in the record look like evolutionary transitions, and therefore the burden of proof falls on anyone who challenges this interpretation. Because the pattern of fossils appears to show life gradually transforming over millions of years, evolutionists claim that the simplest explanation is that evolution actually occurred.

Here is the problem with this argument. Almost all of the fossil record is not like this. Most new forms of life show up abruptly in the fossil record and it appears from the fossil record that most species do not change and stay the same for a very long time. This looks like the exact opposite of evolution taking place. Here are a few references elaborating on this fact:

"The absence of fossil evidence for intermediary stages between major transitions in organic design, indeed our inability, even in our imagination, to construct functional intermediates in many cases, has been a persistent and nagging problem for gradualistic accounts of evolution."[106]

“The extreme rarity of transitional forms in the fossil record persists as the trade secret of paleontology. ... to preserve our favored account of evolution by natural selection we view our data as so bad that we never see the very process we profess to study.”[107]

“No wonder paleontologists shied away from evolution for so long. It seems never to happen. Assiduous collecting up cliff faces yields zigzags, minor oscillations, and the very occasional slight accumulation of changeover millions of years, at a rate too slow to really account for all the prodigious change that has occurred in evolutionary history. When we do see the introduction of evolutionary novelty, it usually shows up with a bang, and often with no firm evidence that the organisms did not evolve elsewhere! Evolution cannot forever be going on someplace else. Yet that’s how the fossil record has struck many a forlorn paleontologist looking to learn something about evolution.” [108]

So the fossil record looks the opposite of what it would look like if evolution was taking place. I have a question though. Based on the version of Occam’s razor that the evolutionists themselves were using, are they going to now accept that the burden of proof is on the person arguing that evolution is taking place? The evolutionists have an ad hoc explanation for the problems their theory faces from the fossil record. This explanation is that the fossil record is poor and does not preserve enough species. But this explanation has massive problems and is rejected even by numerous mainstream paleontologists. Here are two references for this fact:

“In the seven years since Harvard paleontologist Stephen Jay Gould died, there have been only a few assessments of his role in paleobiology and evolutionary theory. Although non-paleontologists still do not realize it, the “punctuated equilibrium” model first proposed by Gould and Niles Eldredge in 1972 has widespread acceptance among paleontologists. Nearly all metazoans show stasis, with almost no good examples of gradual evolution. The most important implication is that fossil species are static over millions of years, even in the face of dramatic climate changes and other environmental selection factors. This widespread stasis cannot be explained by simplistic concepts like "stabilizing selection," and remains a mystery that neontologists have not fully addressed or successfully explained. Despite the hope that paleontology would reach the "High Table" of evolutionary biology, the fact that most biologists largely do not understand or accept the implications of paleontological concepts such as species sorting and coordinated stasis suggests that paleontology still has not made an impact on evolutionary thinking that it deserves."[109]

"Measuring the completeness of the fossil record is essential to understanding evolution over long timescales, particularly when comparing evolutionary patterns among biological groups with different preservational properties. Completeness measures have been presented for various groups based on gaps in the stratigraphic ranges of fossil taxa and on hypothetical lineages implied by estimated evolutionary trees. Here we present and compare quantitative, widely applicable absolute measures of completeness at two taxonomic levels for a broader sample of higher taxa of marine animals than has previously been available. We provide an estimate of the probability of genus preservation per stratigraphic interval, and determine the proportion of living families with some fossil record. The two completeness measures use very different data and calculations. The probability of genus preservation depends almost entirely on the Palaeozoic and Mesozoic records, whereas the proportion of living families with a fossil record is influenced largely by Cenozoic data. These measurements are nonetheless highly correlated, with outliers quite explicable, and we find that completeness is rather high for many animal groups."[110]

Some Evolutionists try to come up with an alternative mechanism for Evolution called Punctuated Equilibrium. There are massive problems with both the logic and the science behind Punctuated Equilibrium. Here is a reference explicitly mentioning that Punctuated Equilibrium was developed to deal with the problem from the fossil record:

"I recount this story at some length, as an introduction to punctuated equilibrium, both because Falconer and Darwin presage in such a striking manner, the main positions of supporters and opponents (respectively) of punctuated equilibrium in our generation, and because the tale itself illustrates the central fact of the fossil record so well — geologically abrupt origin and subsequent extended stasis of most species. Falconer, especially, illustrates the transition from too easy a false resolution under creationist premises, to recognizing a puzzle (and proposing some interesting solutions) within the new world of evolutionary explanation. Most importantly, this tale exemplifies what may be called the cardinal and dominant fact of the fossil record, something that professional paleontologists learned as soon as they developed tools for an adequate stratigraphic tracing of fossils through time: the great majority of species appear with geological abruptness in the fossil record and then persist in stasis until their extinction. Anatomy may fluctuate through time, but the last remnants of a species usually look pretty much like the first representatives. In proposing punctuated equilibrium, Eldredge and I did not discover, or even rediscover, this fundamental fact of the fossil record. Paleontologists have always recognized the long term stability of most species, but we had become more than a bit ashamed by this strong and literal signal, for the dominant theory of our scientific culture told us to look for the opposite result of gradualism as the primary empirical expression of every biologist's favorite subject — evolution itself."[111]

I will explain the logical problem with Punctuated Equilibrium. In both Punctuated Equilibrium and Neo-Darwinism, new traits and features emerge in living organisms due to natural selection acting on random mutations. There are two factors regarding new traits or features emerging in organisms and both of these depend on the overall size of the population.

Evolution or the emergence of new forms of life according to standard evolutionary theory depends on the emergence of new traits via random mutation and those traits becoming established in a population when the organisms having those traits reproduce. The organisms with the new traits reproducing until the traits become widespread are referred to as the trait getting fixed in the population.

Here is the important thing to keep in mind. The bigger the population, the better chance random mutations have to produce beneficial traits. The smaller the population, the better chance the traits have of spreading among the population through reproduction. The fundamental concept behind punctuated equilibrium is that of allopatric speciation.

This concept states that due to an environmental change or geological event, a part of a population of a species gets isolated from the general population of the species. In this small portion of the overall population of the same species, rapid evolution takes place as new traits can get fixed much more easily in a smaller population.

Here is the problem with this concept. In a smaller population, traits can be fixed much more easily. This is true. However, the problem is that in smaller populations, it is much more difficult for random mutations to generate new beneficial traits. Here is a reference about this concept from University of Chicago paleontologist Thomas J. M. Schopf:

"in populations large enough to be reasonably variable, but small enough to permit large changes in gene frequencies due to random drift"[112]

By trying to make the fixing of new traits easier by reducing the population, this mechanism makes the generation of those traits much more difficult. This is the fundamental problem with Punctuated Equilibrium and why changing the population sizes cannot solve the problem, even in principle.

For this and many other reasons, punctuated equilibrium is not widely accepted by evolutionary biologists and is not seen as a solution to the problem of the fossil record. Here are some references regarding this fact:

"The Modern Synthesis (or "Neo-Darwinism"), which arose out of the reconciliation of Darwin's theory of natural selection and Mendel's research on genetics, remains the foundation of evolutionary theory. However, since its inception, it has been a lightning rod for criticism, which has ranged from minor quibbles to complete dismissal. Among the most famous of the critics was Stephen Jay Gould, who, in 1980, proclaimed that the Modern Synthesis was "effectively dead." Gould and others claimed that the action of natural selection on random mutations was insufficient on its own to explain patterns of macroevolutionary diversity and divergence, and that new processes were required to explain findings from the fossil record. In 1982, Charlesworth, Lande, and Slatkin published a response to this critique in Evolution, in which they argued that Neo-Darwinism was indeed sufficient to explain macroevolutionary patterns. In this Perspective for the 75th Anniversary of the Society for the Study of Evolution, we review Charlesworth et al. in its historical context and provide modern support for their arguments. We emphasize the importance of microevolutionary processes in the study of macroevolutionary patterns. Ultimately, we conclude that punctuated equilibrium did not represent a major revolution in evolutionary biology – although debate on this point stimulated significant research and furthered the field – and that Neo-Darwinism is alive and well."[113]

"genetic mechanisms that have been proposed [by proponents of punctuated equilibrium] to explain the abrupt appearance and prolonged stasis of many species are conspicuously lacking in empirical support."[114]

"neither of the contending theories of evolutionary change at the species level, phyletic gradualism or punctuated equilibrium, seem applicable to [explaining] the origin of new body plans."[115]

There is something important to keep in mind. The ad hoc explanations of evolutionists are deeply problematic and should be rejected but even if those were valid, the argument for Evolution is still debunked. The fossil record does not match the predictions that evolutionary theory makes about a gradual, continuous pattern of change over time. Fossils are supposed to show a smooth sequence of intermediate forms connecting ancestral and descendant species, but what we actually observe is a record with long periods of stasis, sudden appearances, and many missing transitions.

Some evolutionists attempt to explain these gaps as the result of an incomplete fossil record, poor preservation, or rapid evolutionary events (punctuated equilibrium). Even if one were to accept these ad hoc explanations as valid, they do not restore the positive evidential force that the fossil record is meant to provide for evolution. The reasoning is simple: the fossil record was presented as direct evidence of gradual evolutionary change. If the record does not display this pattern, then the prediction fails. Accepting patchwork explanations only addresses the gaps; it does not produce the gradual transitions that are supposed to confirm evolution.

In other words, even when evolutionists supply poor and desperate explanations to account for the problems with the fossil record, the positive argument for evolution collapses. The fossil record simply does not provide the continuous evidence that evolutionary theory predicts. At best, these explanations serve as post hoc justifications, leaving the fossil record incapable of functioning as independent, confirmatory evidence of the evolutionary process.

Evolution and God of the Gaps

Vestigial organs are supposedly body parts that once had a function in an ancestral species but do not have a function anymore. Junk DNA are sections of DNA that do not code for protein formation. Therefore, it was assumed that these are functionless sections of DNA. A pseudogene is a segment of DNA that structurally resembles a gene but is not capable of coding for a protein. Pseudogenes are most often derived from genes that have lost their protein-coding ability due to accumulated mutations that have occurred throughout the history of life.

This argument is probably the most common argument that has been made by evolutionists for the past 150 years. Unlike most other arguments, this is used to argue for both life emerging due to blind and unguided processes and for universal common ancestry. The way it is used to argue for common ancestry is that similar functionless things among different animal forms cannot be due to common design and hence are due to common descent.

In this section, I will treat these three arguments as one argument because the same fundamental logic is behind all three arguments. However, I will give ample references and examples supporting my argument regarding all 3 of these things. Here is a reference elaborating on this logic:

"Additionally, there are non-functional genes called pseudogenes that are carried over to successive generations and species. These can act as unique historical markers in the genes that indicate how different animals are connected through a biological lineage (Finlay 2013, 132–193; Futuyma and Kirkpatrick 2017, 345–367). Humans and chimpanzees carry very similar pseudogenes, which further corroborate the claim of humans and chimpanzees having a shared ancestor (Zhang 2014). Thus, there is substantial proof of evolution from genetics."[116]

The problem with this argument is that it relies on the same fallacy evolutionists love to accuse design proponents of: the God-of-the-gaps. As discussed in the second chapter, this fallacy is usually framed as follows. We do not know the physical explanation for a certain phenomenon. Therefore, there is no physical explanation. Therefore, God did it.

The flaw is obvious. Lack of knowledge about a phenomenon does not imply that no natural explanation exists, nor does it automatically imply divine action. Evolutionists frequently accuse theists of using this reasoning when defending miracles, yet their most common arguments for junk DNA, pseudogenes, and vestigial organs employ the very same logic.

Consider the evolutionist argument. We do not know the function or purpose of a section of DNA, a pseudogene, or an organ. Therefore, that thing has no purpose. Therefore, it is useless or functionless. This reasoning has been repeated since Darwin's time. It is a classic God-of-the-gaps argument: ignorance about a structure's function is taken as proof that it is functionless. The irony is striking. While skeptics claim that miracles are unsupported because invoking God is a "gap argument," they themselves build major evolutionary claims on the assumption that unknown functions do not exist.

The funny thing is that we now know that Vestigial Organs, Junk DNA, and Pseudogenes have functions. Here are a few references establishing this fact:

"Vestigial organs have long been a source of perplexity and irritation for doctors and of fascination for the rest of us. In 1893, a German anatomist named Robert Wiedersheim drew up a list of 86 human "vestiges", organs "formerly of greater physiological significance than at present". Over the years, the list grew, then shrank again. Today, no one can remember the score. It has even been suggested that the term is obsolete, useful only as a reflection of the anatomical knowledge of the day. In fact, these days many biologists are extremely wary of talking about vestigial organs at all.

This may be because the subject has become a battlefield for creationists and the intelligent design lobby, who argue that none of the items on Wiedersheim's original list are now considered vestigial, so there is no need to invoke evolution to explain how they lost their original functions. While they are right to question the status of some organs that were formerly considered vestiges, denying the concept altogether flies in the face of the biological facts."[117]

The human genome encodes the blueprint of life, but the function of the vast majority of its nearly three billion bases is unknown. The Encyclopedia of DNA Elements (ENCODE) project has systematically mapped regions of transcription, transcription factor association, chromatin structure and histone modification. These data enabled us to assign biochemical functions for 80% of the genome, in particular outside of the well-studied protein-coding regions. Many discovered candidate regulatory elements are physically associated with one another and with expressed genes, providing new insights into the mechanisms of gene regulation. The newly identified elements also show a statistical correspondence to sequence variants linked to human disease, and can thereby guide interpretation of this variation. Overall, the project provides new insights into the organization and regulation of our genes and genome, and is an expansive resource of functional annotations for biomedical research.[118]

Recent advances have established that the DNA of a pseudogene, the RNA transcribed from a pseudogene, or the protein translated from a pseudogene can have multiple, diverse functions and that these functions can affect not only their parental genes but also unrelated genes. Therefore, pseudogenes have emerged as a previously unappreciated class of sophisticated modulators of gene expression, with a multifaceted involvement in the pathogenesis of human cancer. [119]

Another problem is the flawed analogy evolutionists often use, equating living organisms to designed objects like clocks or televisions. Designers of such objects avoid useless parts, they argue, so vestigial features must indicate a lack of design. But living things are not static objects—they are more like buildings. Architects may design a building with extra rooms that are temporarily unoccupied or repurposed over time. This is not a sign of poor design; it is a prudent design choice.

Junk DNA, pseudogenes, and vestigial organs can be understood in the same way. Many initially presumed "useless" structures had or later acquired functions. The human genome, for example, contains sequences once labeled junk but now known to regulate gene expression and other critical processes. Similarly, organs once dismissed as vestigial have proven functional. These examples show that the evolutionist argument—relying on assumed functionlessness—is logically equivalent to the God-of-the-gaps fallacy they criticize in theistic arguments for miracles.

In conclusion, the argument from junk DNA, pseudogenes, and vestigial organs relies on assuming that what we do not currently understand is functionless—a classic God-of-the-gaps fallacy. Ironically, evolutionists use this logic while criticizing theists for invoking God to explain miracles. History shows that many structures once deemed useless actually have important functions. Unknown functions do not disprove design, just as the existence of miracles does not depend on our ignorance. The selective critique of God-of-the-gaps reasoning reveals a clear double standard in evaluating evidence.

Falsification

The last topic we will cover in this chapter is Falsifiability. One of the most common talking points from Evolutionists is that Evolution is science and Intelligent Design is not science because Evolution is testable and falsifiable whereas Intelligent Design is not testable or falsifiable. This claim from Evolutionists suggests that testability and falsifiability are extremely important in terms of evidence. Here is a reference which summarizes this attitude from Evolutionary Biologists:

“If the hypothesis of intelligent design is not refuted by the fact that organisms have imperfect adaptations, maybe the problem with the hypothesis is that it can’t be refuted by any observation at all. Here we come to a second standard criticism that has been made of creationism: that it is untestable. But what does testability mean? Scientists often answer by using the philosopher Karl Popper’s (1959) concept of falsifiability, thinking that this provides the needed clarification.”[120]

This is often used as a contention against miracles as well. The claim is that miracles should not be believed in because they are not falsifiable. There is a simple problem though. Evolution is not falsifiable and it is very easy to see that it is not falsifiable. I will now get into the specific claim by Evolution that life emerged as a result of unguided physical processes and discuss whether that claim is falsifiable or not.

All modern science assumes the principle of methodological naturalism. Here is a reference explaining this topic:

"Even if all the data point to an intelligent designer, such a hypothesis is excluded from science because it is not naturalistic. Of course the scientist, as an individual, is free to embrace a reality that transcends naturalism."[121]

Here is why this is relevant. According to the Evolutionists, the issue of Design and Evolution has to be settled based on science. I do not think that this issue is entirely rooted in science. I think there are a lot of philosophical assumptions and arguments made by both sides. But this is the assumption of most Evolutionists. According to these same people, Intelligent design can never be proven or accepted by Science, even if evidence is present for it. Keep this in mind.

Now let's return to the falsifiability. The evolutionist claim is that life emerged as a result of blind and unguided physical processes. If this claim is falsified what would be proven true? Intelligent Design. But according to evolutionists that can never happen. Moreover, the only way to prove the evolutionist claim would be to falsify Intelligent Design. But according to evolutionists, that cannot happen either. So according to the criteria the evolutionists themselves are using, their claim is not science and has not and cannot be proven true.

This problem extends to every single argument for Evolution. The claim from them is that similarities are only or mainly due to common descent. Since now we know that there are innumerable similarities not due to common descent, an ad hoc explanation like convergent evolution will be invoked. The fossil record is supposedly gradual and therefore supports evolution. Now that we know it is not gradual, a new ad hoc explanation like punctuated equilibrium saves evolution. Junk DNA being functionless is in line with evolution. Now that we know it has a function, that is also compatible with Evolution. Almost every single argument for evolution has problems like this and for each of them exists at least one and usually a few ad hoc explanations. Skeptics want to whine about miracles not being falsifiable but their own faith in Evolution is not falsifiable.

Here is why this is relevant. According to Evolutionists, the question of Design versus Evolution must be settled strictly on scientific grounds. I do not believe this issue is entirely rooted in science. Philosophical assumptions and reasoning play a significant role on both sides. Yet this is precisely the assumption held by most Evolutionists. According to them, Intelligent Design can never be proven or accepted by science, even if compelling evidence exists. Keep this in mind.

Now, let's return to the question of falsifiability. Evolutionists claim that life emerged through blind, unguided physical processes. If this claim were falsified, what would be proven true? Intelligent Design. Yet, according to Evolutionists, that can never happen. Moreover, the only way to "prove" the evolutionist claim would be to falsify Intelligent Design—but Evolutionists also insist that this is impossible. By their own criteria, then, the evolutionist claim is not scientific and, in fact, cannot be proven true.

This problem extends to virtually every argument for Evolution. For instance, similarities among organisms are said to be evidence of common descent. Yet we now know countless similarities exist that cannot be explained by common descent, prompting ad hoc explanations like convergent evolution. The fossil record is claimed to be gradual, supposedly supporting evolution; when it is revealed to be irregular, ad hoc theories such as punctuated equilibrium are invoked. Junk DNA was once considered functionless, aligning with evolutionary expectations; now that we know it has function, it is claimed to still be compatible with evolution.

Almost every argument for evolution encounters this pattern: anomalies arise, and for each, one or more ad hoc explanations are proposed. Skeptics often complain that miracles are not falsifiable, but the faith of Evolutionists in Evolution itself is equally unfalsifiable.

Conclusion

In conclusion, there is little to no convincing evidence for Evolution. Most of the arguments put forward are riddled with philosophical or scientific problems, and defending them often requires Evolutionists to rely on double standards or distortions. I have not addressed every single argument for Evolution—covering them all would have made this chapter far too long—but I focused on the most persistent and widely cited claims. Frankly, with such meager evidence and even weaker reasoning, it is surprising how this belief has gained such widespread acceptance.

The Moon Landing: A Secular Miracle — evidence and its problems

The Apollo Moon landings of 1969–1972 are often presented as one of humanity's greatest achievements—a triumph of technology, courage, and exploration. For many, the images of astronauts walking on the lunar surface are taken as undeniable proof of human ingenuity. Yet, like other widely accepted narratives, the story of the Moon landing deserves careful scrutiny.

Historical events, no matter how celebrated, should be examined critically. The evidence, the claims, and the circumstances surrounding the Apollo missions raise questions that are seldom addressed openly. From anomalies in photographs and videos to unresolved technical and scientific challenges, the Moon landing story is more complicated than mainstream accounts suggest.

In this chapter, I will explore these questions, examining both the official evidence and the reasons why skepticism is reasonable. By applying the same principles of rational inquiry and burden-of-proof analysis that we used for miracles and evolution, we can assess whether the Moon landing was truly the achievement it is portrayed to be—or if alternative explanations deserve serious consideration.

The historical background and context for the Moon landing

The story of the Moon landing is inseparable from the broader historical and political context of the 1960s. At the height of the Cold War, the United States and the Soviet Union were locked in a fierce competition for technological and ideological supremacy. Space became a symbolic battleground: every achievement in orbit or beyond was a statement of national power, scientific prowess, and political legitimacy. The so-called "Space Race" was not only about exploration—it was about demonstrating dominance on a global stage.

In 1957, the Soviet Union launched Sputnik 1, the first artificial satellite, shocking the world and triggering fears in the United States about falling behind in science, technology, and military capability. The U.S. responded rapidly, creating NASA in 1958 and committing vast resources to space exploration. By the early 1960s, both superpowers had achieved milestones in human spaceflight: Yuri Gagarin became the first man in orbit in 1961, and the Americans followed with Alan Shepard's suborbital flight that same year.

Against this backdrop, President John F. Kennedy delivered his famous 1961 speech declaring the goal of sending a man to the Moon before the decade was out. The speech was a rallying cry for national pride and scientific achievement, but it was also a calculated political statement. The Apollo program that followed—an enormously expensive and complex series of missions—was designed as much to demonstrate American superiority as to achieve scientific discovery.

The technological challenges were staggering. At the time, computers were primitive by today's standards, and the idea of sending humans hundreds of thousands of kilometers into space, landing them safely on the Moon, and returning them to Earth was unprecedented. NASA had to solve problems ranging from life support and navigation to propulsion and radiation exposure, all under intense political and media scrutiny. The pressure to succeed was immense, with billions of dollars invested and national prestige at stake.

This historical context is crucial for understanding the Moon landing narrative. It was not merely a scientific mission; it was a geopolitical statement and a product of its era. The Cold War environment, combined with political ambitions, media spectacle, and technological limitations, created conditions where extraordinary claims—like humans walking on the Moon—would be celebrated as triumphs, sometimes without the rigorous questioning that one might expect for events of such magnitude.

The discussion and my position on the Moon landing

Whenever the Moon landing is questioned, a predictable pattern emerges. Many people who proudly identify as "skeptics," "rationalists," or "pro-science" immediately react with ridicule rather than reason. Instead of addressing the specific problems with the Apollo missions, they frequently resort to a lazy tactic: equating Moon-landing skepticism with belief in a flat Earth. This comparison is not only inaccurate—it is intellectually dishonest.

The shape of the Earth and the reality of the Moon landing are entirely separate questions. One can affirm the roundness of the Earth based on abundant, ongoing, and directly observable evidence: the curvature seen from aircraft, satellite imagery updated daily, circumnavigation, time-zones, star-patterns, and simple experiments available to anyone. These forms of evidence are continuous, contemporary, and reproducible.

The Moon landing, however, is a completely different type of claim. The alleged events supposedly occurred a handful of times between 1969 and 1972, using technology far inferior to what we possess today—then abruptly stopped, never to be repeated for over half a century. Unlike the evidence for the round Earth, the "evidence" for the Moon landing is not constantly observable or independently verifiable. We are asked to accept it primarily on the authority of historical footage, government agencies, and claims made during a time of intense geopolitical pressure.

My own position is straightforward and does not deny technological progress or the reality of space exploration. Unmanned probes and drones have certainly gone to the Moon—both in the past and in the present. This is well-established and consistent with modern engineering capabilities. Furthermore, I believe that a crewed Moon landing could be technologically possible in the future. There is no inherent impossibility in the idea itself.

What I question is simply this: Did humans actually land on the Moon in 1969–1972, with the tools, methods, and conditions claimed by NASA? And on this question, there are far too many unresolved problems, inconsistencies, and suspicious details to simply accept the narrative at face value. When the evidence is examined critically—without appeals to authority or emotional commitments—a very different picture emerges, one that suggests the possibility of fabrication, misrepresentation, or at least a partial cover-up.

Skepticism is not denial. It is a request for sufficient evidence. And in the case of the Moon landing, the burden of proof is much higher than many are willing to acknowledge.

The Burden of Proof

One of the most striking features of the Moon landing debate is how casually many people assume that the burden of proof lies entirely on the skeptic. The mainstream narrative is treated as unquestionable fact, and anyone who expresses doubt is expected to produce overwhelming evidence before the discussion can even begin. This mirrors exactly the same double standards seen in debates about miracles and evolution.

Yet the Moon landing is a historical claim, not a continuously observable phenomenon. It allegedly happened several times more than fifty years ago, under unique political circumstances, using technology that—by today's standards—was primitive. Some try to argue that because it supposedly happened multiple times, it is not a "one-off" event. But this is exactly the kind of flawed reasoning skeptics reject in every other context. A claim being *repeated* does not make it *observable* or *verifiable* today.

People who believe in miracles often attribute multiple miracles to the same person—one saint or one prophet or one spiritual figure performing more than one extraordinary act—but skeptics do not accept those repeated claims as evidence of authenticity. The same standard applies here: whether something is claimed once or five times does not change the fact that all of those claims lie entirely in the unrepeatable past. The Apollo missions, like miracle claims, cannot be revisited, cannot be reproduced in the identical historical conditions, and cannot be verified with present-day observation.

So regardless of the number, the Moon landing still remains a series of past, unrepeatable claims, not a phenomenon we can witness today. With that in mind, the burden of proof cannot simply be offloaded onto skeptics.

The burden of proof does not automatically fall on the person questioning a claim simply because the claim is widely accepted. Popularity is irrelevant. Authority is irrelevant. Institutional prestige is irrelevant. As with any past event that cannot be repeated, tested, or observed today, the burden of proof initially lies with whoever asserts that humans actually traveled to the Moon and returned safely using 1960s technology—technology with orders of magnitude less computing power than a modern smartphone.

If someone insists that the Moon landing is true, they must justify why. Pointing to textbooks, appeals to NASA's reputation, or emotional confidence in "science" does not meet that standard. Nor does mocking skeptics or lumping them together with flat-earthers. That is an evasion, not an argument. The question is not whether space travel is possible in principle—it is whether the specific Apollo missions happened as officially described.

Another common tactic used to shift the burden is to claim that there is "evidence" for the Moon landing—photographs, videos, testimonies, rock samples, and so on—and therefore anyone denying it must refute each piece individually. But this misunderstands the role of burden of proof entirely. The burden must be established *before* evaluating the evidence, not after. Otherwise the discussion becomes circular: "The Moon landing is true because the evidence says so; the evidence is valid because the Moon landing is true."

Even more problematic is the inconsistent standard applied by many self-proclaimed skeptics. When evaluating miracles, they insist that the mere existence of any alternative explanation—no matter how speculative—is enough to dismiss miracle evidence. But when it comes to the Moon landing, the same skeptics refuse to entertain even the possibility of alternative explanations, no matter how plausible or historically grounded they may be. The rule shifts depending on the conclusion they want to defend.

If alternative explanations weaken miracle claims, then by their own logic, alternative explanations must also weaken the Moon landing. If unobserved, unrepeatable events put miracles under suspicion, then the unobserved, unrepeatable Apollo missions should be treated the same way. If institutional trust is insufficient for religious claims, then institutional trust cannot be sufficient for NASA either.

The reality is that the burden of proof is shared equally by anyone making a claim—whether affirming or denying. A fair standard of rational inquiry does not begin with assumptions about who is right; it begins with consistency. And when that consistency is applied, the Moon landing is subject to the same scrutiny as any other past claim that cannot be revisited or reproduced. It must be justified, not assumed.

The motivation for fabrication

Some argue that the United States government would have had little reason to fabricate the Moon landing, but a closer look at the historical circumstances reveals powerful incentives for deception—more than enough to motivate a staged "victory." In the 1960s, America was struggling on multiple fronts: fulfilling John F. Kennedy's bold promise to land a man on the Moon had become a national obsession, and after his assassination, achieving that goal carried the emotional weight of legacy as much as politics. Failing to deliver would have been deeply embarrassing, both for the nation and for Kennedy's memory.

Meanwhile, the Cold War raged on. The Space Race was not simply about exploration, but about symbolic dominance: outpacing the Soviet Union mattered not just for science, but for global prestige. If America could convincingly claim a Moon landing, this would register worldwide as proof of power, technological supremacy, and ideological superiority. Tied to this was the arms race: demonstrating mastery in rocketry and guidance systems had strategic military implications, even if the actual landing was illusory.

At home, the United States government was also under significant strain. The Vietnam War was increasingly unpopular, draining resources and dividing public opinion. A televised, triumphant Moon landing would serve as a powerful distraction, a unifying spectacle to bolster national pride and shift the narrative away from war and dissent. For a government under duress, the political payoff of fabricating such an event could have been enormous.

If one accepts motivation as a legitimate reason to doubt miracle claims, then the same standard must apply here. Skeptics who argue that prophets might lie (or exaggerate) for their own ends must confront the fact that the U.S. government was institutionally well-equipped to lie on a mass scale—and had precedents. The record of modern governments is full of secrecy, covert operations, and propaganda. It is not far-fetched to imagine a powerful state executive concocting or embellishing a narrative to achieve political goals.

Contrast this with the early Islamic community. We have strong historical evidence that many of the earliest Muslims displayed unusual transparency and sincerity, even when it was against their own interest. For example, early transmitters of hadith did not shy away from recording and transmitting reports that were not in their interest to transmit—such as Hadiths praising Ali ibn Abi Talib RA and his family, despite the political tensions associated with his family. Their commitment to recording and preserving tradition was not limited to flattering stories, but included all manner of narrations regardless of factional leanings.

Another powerful example comes from the aftermath of the Muslim conquest of Syria. According to early Islamic sources and even Christian sources, when the military governor Abu Ubaydah ibn al-Jarrah RA recognized that his forces would be unable to defend certain cities that they had conquered, he ordered that all the jizya previously collected from the non-Muslim inhabitants be returned. He stated that the agreement with those communities was contingent on protection—a promise he now could not keep—and thus he returned their payments. Historical records note that large sums were refunded and that Christians rejoiced and prayed for him, saying that if the Roman (Byzantine) authorities had remained in power, they would never have returned their money. Here is an early Christian source for this incident:

Following the death of the Prophet Muhammad, in the reign of the second Caliph, Umar bin Khattab, the Muslim armies began liberating the people of Syria from the Byzantines (Romans). The Christians of Syria were divided into many different denominations, such as Monophysites, Jacobites and Nestorians. Almost all of them were facing severe persecution at the hands of the ruling Byzantine Church. Dionysius of Tel-Mahre, a Jacobite patriarch from 818 to 845 CE, stated in his chronicle that the Byzantine Emperor Heraclius sent an army to expel the Muslims out of Syria and recapture the land. The Muslim forces decided to withdraw from Syrian cities in order to fight an open pitch battle with the Byzantines. Whilst pulling back, the Muslims decided, out of fairness, to refund the money which they had taken as a tribute from the Syrian Christians.[122]

This is not a case of self-serving propaganda: it is an instance of publicly recorded financial restitution, under conditions of military risk, documented by early Muslim historians. The fact that Muslims returned money when they could not fulfill their obligations speaks to a standard of integrity and contract-based governance.

Given this historical contrast, the idea that the U.S. could have fabricated a Moon landing is not only possible but intelligible—and perhaps even likely, given the incentives. If skeptics demand proof of sincerity or motive in miracle claims, they must also confront motive and the capacity for deceit when evaluating state-sponsored narratives. Dismissing the latter while elevating the former betrays philosophical inconsistency and selective skepticism.

The possibility of alternative explanations and conspiracies

When evaluating the Moon landing, it is important to recognize that multiple alternative explanations for the event exist, even if we do not assert any of them as definitive. One plausible scenario is that the U.S. government could have created the appearance of a successful lunar mission primarily as a political and financial strategy, rather than a fully realized technological achievement. The Apollo program, after all, involved immense sums of money—billions of dollars in 1960s funding—and the government had strong incentives to justify this expenditure to Congress, the public, and international observers. In such a context, it is not inconceivable that appearances could have been prioritized over actual results, particularly if doing so enhanced national prestige and bolstered support for the administration.

Historical examples suggest that the U.S. government has, on multiple occasions, misrepresented or manipulated programs for strategic purposes. The Vietnam War, for instance, saw repeated misstatements of military progress and casualty figures to maintain public support. Similarly, the Iran-Contra affair revealed that the government could covertly fund operations while misleading both the public and Congress. The 1950s and 1960s also saw exaggerated claims about progress in the arms and space races, with NASA itself facing internal pressure to meet deadlines and political expectations, sometimes stretching technical realities to present optimistic narratives. These precedents show that large organizations, including the U.S. government, have a track record of presenting appearances that serve political or financial ends.

Applying this perspective to the Moon landing, one can imagine a scenario in which the program's achievements were partially staged or presented in a way that maximized public confidence, without necessarily reflecting the full reality of the mission. It is not necessary to claim that this *definitely* happened; the point is that such an explanation is plausible based on historical patterns. Yet skeptics who demand rigorous proof before accepting miracles would insist that any such alternative explanation must be definitively disproven before the original event could be considered credible. This reveals an inconsistency: if the possibility of deception, misrepresentation, or mismanagement exists, it should weaken confidence in the Moon landing by the same logic that skeptics apply to miracles.

The widespread acceptance by other governments

The widespread acceptance of the moon landing by other governments is often cited as one of the strongest arguments for its occurrence. At first glance, it seems compelling: if multiple countries, including powerful and rival nations, acknowledged that the United States had landed on the moon, it would appear unlikely that the event was fabricated. Among these, the Soviet Union's acceptance is particularly striking. After all, during the Cold War, the USSR and the USA were engaged in intense technological, military, and ideological rivalry. Skeptics often argue that if the moon landing were falsified, the Soviets would have exposed it immediately to discredit the United States.

However, this argument has notable complications. The acceptance of the moon landing by other governments was not the result of independent verification in every case. Many countries were heavily influenced by one superpower or the other. Nations under U.S. influence would likely accept the official narrative, while those aligned with the USSR would echo Soviet positions. In such circumstances, acceptance often reflects geopolitical alignment rather than genuine verification.

Even the USSR's acknowledgment, while impressive, is not without alternative explanations. It is plausible that the Soviet government, facing other strategic priorities, found it expedient to accept the U.S. claim in exchange for political, technological, or economic concessions. During the Cold War, both the United States and the Soviet Union engaged in multiple instances of tacit cooperation, coordinated silence, or mutually beneficial cover-ups. For example, during the Cuban Missile Crisis in 1962, both sides initially withheld full disclosure of nuclear movements to prevent public panic and escalation, while negotiating in secrecy. The Partial Test Ban Treaty of 1963 involved mutual limitations on nuclear testing, but both sides sometimes delayed or misrepresented the scope of compliance for strategic advantage. Similarly, both nations covertly withheld or manipulated information about nuclear accidents, espionage missions, and satellite reconnaissance to avoid public scandal or international tension. Another example is the 1972 SALT I negotiations, where both countries deliberately underreported certain missile developments while negotiating arms limitations, creating a tacit coverup that benefited both.

Thus, while the USSR's acceptance of the moon landing is a strong point in favor of its historicity, historical precedent shows that rival governments can, under strategic or political pressure, agree to remain silent or manipulate the public narrative. The existence of such coordinated or mutually beneficial coverups demonstrates that acceptance by a rival government, though significant, is not definitive proof of the event itself.

The Moon rocks

One of the commonly cited arguments in favor of the moon landing is the existence of lunar rocks brought back by the Apollo missions. These rocks are said to have distinctive mineral compositions, isotopic ratios, and features that differ from typical Earth rocks, which has led many to argue that their existence strongly supports the historicity of the landings. The argument is straightforward: the rocks could not have originated on Earth, so they must have been collected from the Moon.

However, there are alternative explanations that must be considered. Some lunar rocks share surprising similarities with Earth rocks, which has led skeptics to suggest that at least some samples could be terrestrial in origin or even meteorites that naturally arrived from the Moon via asteroid impacts. Indeed, hundreds of lunar meteorites have been found on Earth over the decades, providing a naturally occurring source of moon-like material. If one were to apply the same reasoning to miracles, one could argue that evidence may exist in multiple forms yet still be misattributed or coincidental.

Another possibility is forgery or misidentification. Historical cases demonstrate that not every "moon rock" presented as genuine has withstood scrutiny. For example, the Netherlands received a supposed moon rock in 1969, a gift from the U.S. Ambassador to the former Prime Minister Willem Drees. Later analysis revealed this rock to be petrified wood, not lunar material. Such instances show that errors, misrepresentation, or even intentional substitution can occur, which complicates using moon rocks as proof of the Moon landing.

In conclusion, while lunar samples brought back by the Apollo missions present strong evidence supporting the moon landings, alternative explanations—similarity to Earth rocks, natural lunar meteorites, and misidentifications—exist and must be acknowledged. The possibility of deliberate forgery or error, however rare, is also a factor that skeptics can legitimately raise, much as one might with historical claims of miracles. This does not prove a conspiracy, but it highlights the need for careful scrutiny before accepting any claim as absolute evidence.

The Retroreflectors placed on the Moon

One of the arguments often presented in favor of the Moon landing is the presence of retroreflectors—special devices designed to reflect laser beams sent from Earth. Apollo 11, 14, and 15 missions all deployed these reflectors, which allow scientists to measure the exact distance to the Moon by timing how long it takes for a laser beam to bounce back. Proponents of the Moon landing argue that their existence is strong evidence that humans physically visited the Moon.

However, the Moon naturally reflects sunlight, and this diffuse reflection was detectable even before the Apollo missions. In 1962, just seven years before Apollo 11, astronomers successfully bounced laser beams off the lunar surface, though the return signal was weak and scattered. This shows that some observed reflections could originate from natural properties of the lunar surface, and it is therefore plausible that areas identified as retroreflectors could simply be naturally reflective regions rather than devices placed by astronauts.

Another equally plausible explanation is that the retroreflectors were placed by unmanned robotic missions. The Soviet Luna program, for instance, successfully landed probes such as Luna 17 and Luna 21, which carried instruments to the lunar surface, demonstrating that non-human spacecraft could deploy devices or leave detectable artifacts. These missions predate and overlap with the Apollo era, showing that human presence is not the only way reflectors could have been placed.

If this were an argument supporting a miracle, and alternative explanations existed that were just as plausible—such as natural properties or automated placement—skeptics would demand that proponents rule them out before accepting the claim. The retroreflectors, therefore, cannot be treated as conclusive proof of manned lunar landings without careful consideration of these reasonable alternatives.

The problems with the footage

We now turn to some of the arguments that skeptics raise against the Moon landing, particularly issues with the photographic and video footage. Two of the most commonly cited anomalies are the movement of the American flag and the appearance of perpendicular or inconsistent shadows in lunar photographs.

The American flag planted on the Moon appears to wave or ripple in several videos. Skeptics highlight this as a problem because there is no atmosphere on the Moon—no air to cause motion. NASA's official explanation is that the flag moves when it is being twisted and planted by the astronauts, and that the ripples are simply a result of the flag's material and the folds it retained when stored. While this explanation is physically possible, it raises questions: the flag continues to appear to move even after the astronauts stop touching it, and some frames show the fabric appearing to wave as if blown by wind. To account for this, proponents must rely on inertia and subtle vibrations from the astronauts' actions, which can seem implausible to casual observers, especially given the expectation that the flag should hang motionless in a vacuum.

Another issue is the appearance of perpendicular or non-parallel shadows in lunar photographs. Shadows that diverge or converge unnaturally suggest to skeptics the presence of multiple light sources, leading to claims of studio lighting. The official explanation is that the uneven lunar surface and the perspective of the camera create optical effects that make shadows appear non-parallel, even though the Sun is the only light source. While this is physically accurate—uneven terrain can distort shadows—skeptics argue that the precise combination of terrain angles, camera placement, and sunlight required to produce these effects seems unlikely. The claim rests on a subtle understanding of optics and perspective, which is not immediately intuitive, giving the appearance of inconsistency.

In both cases, the responses rely on explanations that are physically possible but require specific, sometimes improbable conditions. The astronauts' handling of the flag must produce just the right motion, and the terrain must distort shadows in exactly the ways observed. Skeptics point out that these "solutions" are post-hoc interpretations designed to fit the footage, rather than evidence that the footage is definitively genuine.

The deeper point here is one of standards. When skeptics examine miracles, they often reject plausible alternative explanations, insisting on near-impossible proof before acknowledging the phenomenon. Yet in the case of the Moon landing, equally implausible explanations—such as careful camera angles and the flag's inertia—are accepted as sufficient to resolve anomalies. This highlights a double standard: similar criteria for plausibility are applied inconsistently depending on whether the claim supports the dominant narrative or challenges it.

The lost and destroyed equipment and footage

After examining the problems with the footage, another significant point often raised by skeptics concerns the original materials themselves. The original Moon landing tapes—the highest-quality recordings made directly from the slow-scan television signals received from the lunar surface—were reportedly erased or lost by NASA in the 1980s.[123] Additionally, much of the original equipment used during the Apollo missions, including some critical instruments, telemetry tapes, and even engineering models, was either discarded or destroyed over time.[124] This means that modern investigators cannot directly access the pristine, primary sources and must rely on copies or secondary recordings that have undergone multiple generations of reproduction, with possible degradation or alteration.

From a reasoning perspective, this situation is strikingly similar to what happens when evidence for a claimed miracle is destroyed or unavailable. In both cases, the inability to examine the original sources does not automatically prove that the event occurred or did not occur. However, skeptics of miracles often treat the loss of evidence as a fatal blow to the claim, demanding near-impossible verification before acceptance. If the same standards were applied consistently, the loss or destruction of original Apollo materials should, at minimum, warrant serious caution in treating the Moon landing as incontrovertible, rather than assuming its absolute certainty. This raises the question of double standards: why is destroyed evidence a problem for miracles but not for the Moon landing?

The problems regarding the Van Allen belt radiation

The Van Allen radiation belts are often cited by skeptics as a major obstacle to the Moon landing. These belts contain highly energetic charged particles trapped by Earth's magnetic field, and the skeptics argue that exposure to such intense radiation would have been lethal for astronauts without unprecedented shielding. According to this view, passing through the belts would have been physically impossible with 1960s technology, making the Moon landing highly implausible.

Proponents of the Moon landing counter this argument by pointing out that the Apollo missions passed through the belts quickly, minimizing exposure. They also emphasize that the spacecraft had some shielding and that calculations suggested the total dose received by astronauts was within survivable limits. On the surface, this appears to address the concern. However, this is just an appeal to blanket possibility. The proponents could, in principle, simulate the conditions of the Van Allen belt and have people stand in them or pass through them for the same duration that the Apollo astronauts did, but they are not doing this.

Moreover, one does not even need to go all the way to the Moon to pass the Van Allen belt. Simply passing through the belts would be enough to demonstrate the feasibility of surviving that level of radiation. Yet no one after the Apollo missions has even attempted to do this to prove that it is possible.

This is analogous to the demand for evidence in miracle claims. Skeptics often ask for demonstrable proof before accepting extraordinary events. In the case of passing through that level of radiation, proponents assert it is possible but offer only theoretical plausibility rather than independent, reproducible demonstration. Appeals to possibility cannot substitute for direct confirmation, and the absence of attempts to verify these conditions raises questions about the strength of the claim.

The requirement of conspiracy

Proponents of the Moon landing often argue that the event could not have been faked because doing so would require a massive, global conspiracy involving hundreds of thousands of engineers, scientists, technicians, astronauts, media personnel, and foreign observers. The claim is that the scale of such deception would be impossible to coordinate and even more impossible to keep secret—especially over decades.

This argument sounds strong at first, but it rests on an exaggerated and unrealistic picture of how real conspiracies work. You do not need "millions of people" to knowingly participate in a fabrication. You only need the small group at the top—the people with full access, decision-making authority, and control over information. Everyone beneath them can be compartmentalized, given only partial tasks, told only what they need to know, and kept unaware of the wider picture. Military and intelligence agencies operate this way by design. Most complex government programs already involve strict compartmentalization, classified layers, and need-to-know structures. It is entirely normal for thousands of people to work on a project without any understanding of the final product.

History repeatedly shows that small groups of powerful individuals can orchestrate and conceal major wrongdoing even when countless people are indirectly involved. The Jeffrey Epstein scandal is a modern example: a single individual, with the help of a small network of elites and officials, was able to traffic minors, manipulate legal outcomes, hide evidence, and escape serious accountability for years—despite the involvement of pilots, staff, security, hotels, politicians, and agencies that interacted with him without understanding the whole picture. And even after his death, the full truth remains obscured, with powerful institutions still refusing transparency. If such a crime, involving high-profile figures and enormous public interest, can be covered up by a small circle, then the idea that a Cold War–era space program—controlled by the U.S. government, shielded by national security, and driven by geopolitical stakes—could maintain secrecy is hardly impossible.

Thus, the argument that the Moon landing could not have been faked because "too many people would have to be in on it" fails under even basic scrutiny. Real conspiracies are maintained not by involving everyone, but by ensuring that only a handful know the truth while everyone else performs their role without full understanding. The Moon landing would be no different.

<u>The supposed impossibility of faking the Moon landing</u>

Proponents of the Moon landing often claim that faking the missions would have been impossible. They argue that the photographs, videos, and live broadcasts show details of the lunar surface, lighting, and astronaut behavior that could not possibly have been reproduced on Earth with 1960s technology. To them, the footage is so convincing that it proves the missions truly happened. This line of reasoning suggests that the "impossibility" of faking the Moon landing works as evidence for its authenticity.

However, this argument fails under closer examination. The essential point is that the difficulty of producing realistic lunar footage does not compare to the difficulty of actually sending humans to the Moon. Even in the 1960s, it was possible—though expensive and time-consuming—to simulate environments convincingly with controlled lighting, scaled models, and practical effects. Today, with modern technology, creating such simulations is even easier and cheaper. In contrast, sending humans to the Moon required massive rockets, life-support systems, precise navigation, and enormous funding. It remains vastly more difficult and costly than creating convincing footage.

The comparison highlights a critical flaw: the argument assumes that technical difficulty is a reliable proxy for authenticity. But we cannot say "the footage is real because it would have been too hard to fake," when in reality, faking it is far easier than accomplishing the actual mission. This undermines the standard claim that the "impossibility of faking" proves the Moon landing, showing instead that the reasoning is based on a misconception of relative difficulty.

The biggest problem with the Moon landing

The biggest problem with the Moon landing narrative is the fact that, despite half a century of immense technological progress, humanity has been unable to return to the Moon. If the official story is true, then the technological pattern of the Apollo era stands as a unique historical anomaly: a period in which we supposedly achieved something vastly more difficult than anything accomplished afterward, only to permanently lose the ability to do it. This contradicts the entire observed progression of technology, which overwhelmingly becomes more reliable, more efficient, and significantly cheaper over time. In every other field of human achievement—aviation, computing, medicine, engineering—we see consistent, predictable advancement. There are no comparable cases where society accomplished something extremely advanced, then regressed for decades, only to struggle even to recreate what was done before with weaker tools.

Even if humans eventually return to the Moon in the coming years, this still would not explain the unprecedented 50-year inability to repeat what was allegedly routine in the late 1960s and early 1970s. A future Moon landing does not retroactively justify a half-century of incapacity. If anything, it highlights the gap even more: why was a task supposedly possible with 1960s technology impossible for the far superior technology of the 1980s, 1990s, 2000s, 2010s, and early 2020s?

The financial explanation often offered by proponents only deepens the difficulty. Cost should decrease over time due to technological advancements, not increase. The United States in the 1960s was simultaneously funding the Vietnam War, engaged in an expensive Cold War, and possessing computational power millions of times weaker than today. Yet it supposedly afforded the Moon with ease, while governments with vastly stronger economies and technology have found it financially out of reach. This makes little sense. If Apollo was truly achieved under the constraints of the 1960s, then repeating it with modern tools, robotics, materials, and computation should be dramatically easier and cheaper. The fact that it is not reinforces the suspicion that the original achievement did not occur as claimed.

Compare this with the strongest pro-Moon landing argument—the USSR's acceptance of the landing. In that case, there are clear precedents and plausible scenarios in which geopolitical rivals cooperate, stay silent, or trade concessions behind the scenes. We have real-world examples of Cold War cover-ups and intelligence agreements. But when it comes to technological regression on this scale, there are no counterexamples at all. We do not see any other case of humanity accomplishing an extraordinary technological feat and then losing the ability to repeat it for generations. The absence of such parallels makes this a stronger argument than any geopolitical explanation.

In short, the timeline of human technological advancement does not support the idea that a civilization goes to the Moon early in its spacefaring history, then loses the capability entirely for half a century despite unprecedented progress. This contradiction stands as the single largest problem with the Moon landing narrative.

Conclusion

In the end, the question of the Moon landing is not about denying human achievement, nor about embracing wild speculation. It is about applying a single, consistent standard of reasoning. When we follow that standard, the official narrative raises far more unresolved problems than most people realize. The footage is questionable, key evidence has been lost or destroyed, the physical challenges remain untested, alternative explanations exist, and the most powerful argument in its favor—the USSR's acceptance—still has plausible counter-scenarios. Above all, the inability to return to the Moon for over half a century, despite vastly superior technology, is an anomaly that does not fit any ordinary pattern of technological progress.

None of this proves with absolute certainty what did or did not happen. But it does show that blind confidence in the Apollo story is no more rational than blind confidence in any other unexamined belief. A fair and honest approach requires either confronting these issues directly or admitting that the matter is far from settled.

Ad Hoc Explanations and Selective Skepticism — parallels in all three

Having examined the evidence for and against miracles, Evolution, and the Moon landing, one striking pattern becomes clear: when a claim cannot be directly observed or repeated in the present, defenders and skeptics alike often rely on ad hoc explanations. These are improvised rationalizations—designed less to illuminate reality than to preserve a preferred conclusion.

All three cases share this challenge: none are happening right now. We cannot witness past evolutionary transitions unfolding, watch a historical Moon landing live, or observe ancient miracles firsthand. The absence of direct, repeatable observation creates a vacuum that must be filled. Ad hoc explanations appear to bridge that gap—whether it's dismissing inconvenient evidence, reinterpreting testimonies, or inventing mechanisms after the fact.

In this chapter, we will explore the primary ad hoc explanations for Evolution, the Moon landing, and miracles, as well as those used in arguments for or against them. By highlighting these parallels, we can see how selective skepticism—doubt applied inconsistently—shapes debates across science, history, and theology. Recognizing these patterns is essential for fair and consistent reasoning about events that cannot be observed directly.

The ad hoc explanations regarding Evolution

One of the most common responses from evolutionists when confronted with the challenge that large-scale evolutionary changes are not observable today is to point to small, ongoing changes in living organisms. Indeed, minor adaptations—such as bacteria developing resistance to antibiotics or variations in beak shapes among birds—can be directly observed. The problem with this response is that such small changes were also visible long before Darwin formulated his theory. What remains unobserved, however, are the major transitions Evolutionists claim have occurred over millions of years.

When evidence of large-scale evolutionary change is absent, evolutionists often resort to an ad hoc explanation: evolutionary processes are simply too slow to witness within a human lifespan. According to this claim, millions of years are required for major changes, making them invisible in the present. At first glance, this seems reasonable—but closer examination reveals multiple problems.

Firstly, there is very little evidence supporting the notion that evolutionary processes must be slow. The primary evidence for Evolution—such as similarities in the anatomy, genetics, and embryology of living organisms—demonstrates common descent, not the speed of change. Likewise, the few examples of small, observable changes do not indicate that large changes necessarily require millions of years.

Secondly, the fossil record—the only potential evidence for slow change—does not consistently support gradualism. As discussed in the previous chapter, fossils often appear suddenly, with fully formed forms, rather than through a slow, continuous process. Evolutionists therefore must invent additional ad hoc explanations to reconcile this discrepancy, claiming that the record is incomplete or that intermediate forms were unlikely to fossilize. It is because of the problems with the fossil record that Evolutionists have to come up with ad hoc explanations like Punctuated Equilibrium which states that Evolution is too fast to be observed. We discussed the problem with this earlier but it can be easily seen that these are just ad hoc explanations made up to handwave away the problems facing Evolution.

Ultimately, the claim that Evolution is "too slow to observe" functions as a blanket ad hoc justification. It is not grounded in independent evidence, but merely serves to preserve the theory in the face of absent large-scale observations. By relying on this assumption, evolutionists shift the debate away from empirical scrutiny and into the realm of untestable possibilities.

Another common ad hoc explanation used by evolutionists concerns the fossil record. Since Darwin's time, it has been repeatedly argued that the apparent gaps in fossils—especially the lack of transitional forms—do not disprove Evolution, because these intermediate fossils are supposedly "missing" or unlikely to have been preserved. This justification has been a standard response whenever the fossil evidence seems incomplete or contrary to expected gradual changes.

The problem with this explanation is that it is increasingly challenged even within mainstream paleontology. Modern research suggests that the fossil record is far more complete and reliable than the traditional "missing fossils" argument allows. Many transitional forms have been discovered, and patterns in the record—such as the sudden appearance of new species—pose difficulties for the gradualistic model without needing to invoke hypothetical missing data. Here are some references regarding this fact:

"Measuring the completeness of the fossil record is essential to understanding evolution over long timescales, particularly when comparing evolutionary patterns among biological groups with different preservational properties. Completeness measures have been presented for various groups based on gaps in the stratigraphic ranges of fossil taxa and on hypothetical lineages implied by estimated evolutionary trees3. Here we present and compare quantitative, widely applicable absolute measures of completeness at two taxonomic levels for a broader sample of higher taxa of marine animals than has previously been available. We provide an estimate of the probability of genus preservation per stratigraphic interval, and determine the proportion of living families with some fossil record. The two completeness measures use very different data and calculations. The probability of genus preservation depends almost entirely on the Palaeozoic and Mesozoic records, whereas the proportion of living families with a fossil record is influenced largely by Cenozoic data. These measurements are nonetheless highly correlated, with outliers quite explicable, and we find that completeness is rather high for many animal groups."[125]

"Broadly, however, the astonishing agreement between evolutionary trees based on molecular (DNA) data and the fossil record, and the fact that the detailed pattern of the evolution of life has not changed much over 200 years of study, suggests that the fossil record is adequate for the purpose we have put it to. It provides incontrovertible evidence of Darwinian evolution. And it can still supply surprise, astonishment and wonder."[126]

"if scaled to the ... taxonomic level of the family, the past 540 million years of the fossil record provide uniformly good documentation of the life of the past."[127]

"Morphological analysis of four higher taxa of fossil marine invertebrates shows that, over the history of paleontology, there is no general tendency for morphologically extreme or modal species and genera to be described preferentially early or late..... Although we still have much to learn about the evolution of form, in many respects our view of the history of biological diversity is mature."[128]

"Our exaggerated emphasis on the imperfection of the fossil record feeds the perception among scientists in general that the fossil record is an unusually poor data set. It isn't.....

Through her comprehensive examinations of live-dead comparisons, Susan Kidwell (2002, 2013) showed the fossil record contains a high-fidelity record of species richness and especially abundance, a pattern both unexpected and most welcome."[129]

"Although the fossil record preserves a wealth of historical data about past ecosystems, the current paradigm, which postulates that fossils provide faithful archives of ecological information, stems from research primarily focused on a single group of organisms known for their high fossilization potential: molluscs. Here, we quantify the fidelity of higher taxa (six phyla and 11 classes) by comparing live communities and sympatric dead remains (death assemblages) using comprehensive surveys of benthic marine invertebrates from coastal habitats in North Carolina (U.S.A). We found that although community composition differed between the two assemblages across phyla and classes, these differences were predictable with an overabundance of robust and more preservable groups.

In addition, dead molluscs appear to be an excellent proxy for all taxa when tracking spatio-temporal patterns and shifts in community structure using a variety of ecological metrics, including measures of α, γ, and β diversity/evenness. This suggests that despite filters imposed by differential preservation of taxa and time-averaging, the fossil record is likely to be reliable with respect to relative comparisons of composition and diversity in shallow benthic marine paleocommunities. This is consistent with previous work indicating that shallow marine death assemblages can yield robust ecological estimates adequate for assessing the variability of ecosystems that existed under natural, pre-anthropogenic conditions."[130]

Thus, the "missing fossils" argument functions primarily as an ad hoc rationalization. It shifts attention from actual evidence to hypothetical scenarios that preserve the theory, rather than confronting the empirical record directly. While this explanation has historical roots, its continued use today reflects a reliance on improvisation rather than on demonstrable data.

A third, often unacknowledged, ad hoc explanation concerns convergent evolution—the phenomenon in which similar traits arise independently in species without a common ancestor. As explained earlier, the core argument for Evolution relies heavily on the similarities between living organisms. Evolutionists claim that these similarities are unlikely without common descent. By that logic, convergent evolution—similarities arising without shared ancestry—should be rare and exceptional.

The problem is that convergent evolution is actually widespread across the tree of life. Yet rather than acknowledging that this undermines the logic of their main argument, evolutionists handwave the issue and claim that convergent evolution is, in fact, a "prediction" of Evolutionary theory. This reinterpretation is not based on prior evidence or independent reasoning—it is a post hoc rationalization designed to protect the theory from contradiction.

In other words, what should have been a significant challenge to the evolutionary framework is instead treated as a confirmation, in direct contradiction to the original argument. As the logic of Evolutionary reasoning shows, the prevalence of convergent traits directly conflicts with the claim that similarities necessarily indicate common descent. Calling this a fulfilled prediction is therefore a classic ad hoc maneuver: it preserves belief in the theory by redefining or handwaving away evidence that contradicts it. Here are some references which show that widespread convergent evolution is not something that was expected based on the evolutionary theory:

"Charles Darwin found the convergent evolution of electroreception in the electric fishes to be so unusual and so improbable that he included it in On the Origin of Species in his list of difficulties for the theory of natural selection to explain: " The electric organs offer another and more serious difficulty; for they occur in only about a dozen dishes, of which several are widely remote in their affinities. Generally when the same organ appears in several members of the same class, especially if in members having very different habits of life, we may attribute its presence to inheritance from a common ancestor; and its absence in some of the members to its loss through disuse or natural selection. But if the electric organs had been inherited from one ancient progenitor thus provided, we might have expected that all electric fishes would have been specially related to each other " (Darwin 1859, 193). Today we know that the convergent distribution of electrosensory organs in fish is much more widespread than Darwin realized, and that they have convergently appeared even in the mammals. Rather than posing a difficulty for the theory of natural selection, the convergent evolution of electrosensory organs in distantly related animals is seen today as a prime example of functional constraint and the process of natural selection in action (Zakon and Unguez 1999, Hopkins 2008)."[131]

"So, too, it is now widely thought that the history of life is little more than a contingent muddle punctuated by disastrous mass extinctions that in spelling the doom of one group so open the doors of opportunity to some other mob of lucky-chancers. The innumerable accidents of history and the endless concatenation of whirling circumstances make any attempt to find a pattern to the evolutionary process a ludicrous exercise. Rerun the tape of the history of life, as S. J. Gould would have us believe, and the end result will be an utterly different biosphere. Most notably there will be nothing remotely like a human, so reinforcing the notion that any other biosphere, across the galaxy and beyond, must be as different as any other: perhaps things slithering across crepuscular mudflats, but certainly never the prospect of music, nor sounds of laughter. Yet, what we know of evolution suggests the exact reverse: convergence is ubiquitous and the constraints of life make the emergence of the various biological properties very probable, if not inevitable."[132]

In all three cases—small observed changes, missing fossils, and convergent evolution—we see evolutionists relying on ad hoc explanations to reconcile the theory with reality. These justifications are invoked precisely because the large-scale evolutionary changes predicted by the theory are not occurring before our eyes. Where direct observation falls short, appeals to slowness, hypothetical gaps, or reinterpretations of convergent traits are used to preserve the narrative. This pattern reveals a consistent reliance on post hoc reasoning rather than empirical verification, highlighting a fundamental tension between the theory's claims and the evidence available in the present.

The ad hoc explanations regarding the Moon landing

The central issue requiring ad hoc explanations in the Moon landing narrative is simple: it is not happening right now. Unlike phenomena that can be observed and repeated, such as natural processes or contemporary technological feats, the Moon landing allegedly occurred between 1969 and 1972 and has not been repeated. This absence creates a gap between claim and ongoing observation, which necessitates ad hoc reasoning to reconcile.

The primary ad hoc explanation offered is that returning humans to the Moon today is prohibitively expensive. According to this argument, financial and logistical constraints prevent repetition, explaining why no modern mission has matched Apollo's achievements. Yet this reasoning raises questions: technological progress tends to reduce cost and increase reliability over time. Modern spacecraft, computers, robotics, and materials vastly exceed 1960s capabilities, making the "too expensive" claim difficult to justify without relying on speculative assumptions.

This explanation mirrors the logic of ad hoc reasoning in other domains, such as evolution and miracles: it appeals to a possibility (here, financial and logistical limits) without providing direct evidence. Just as evolutionists claim that evolutionary processes are "too slow to observe" or miracle skeptics rely on human deception, Moon landing proponents must invoke contingencies that cannot be independently verified to preserve the plausibility of their claim.

Ultimately, the reliance on such ad hoc reasoning highlights the key challenge: the Moon landing is a past, unrepeatable claim, not a continuously observable phenomenon. Financial constraints, technological regressions, or other ex post facto explanations are introduced not because they are empirically verified, but because they are necessary to reconcile the narrative with the simple fact that no one is currently walking on the Moon.

The ad hoc explanations regarding Miracles

A common objection raised by skeptics is: *"If miracles are real, why aren't they happening right now?"* This objection assumes that miracles should function like technological processes—repeatable, ongoing, and available on demand. But within Islam, this assumption is misplaced. Islam does not teach that miracles would continue occurring regularly after the time of Prophet Muhammad ﷺ. Rather, the primary purpose of miracles in Islam is to authenticate a prophet's truthfulness. Since the Prophet Muhammad ﷺ is the final prophet, there is no theological reason to expect prophetic-level miracles to be happening today. Unlike phenomena such as space travel—where advancing technology should lead to more frequent demonstrations—miracles are not part of a continuous natural process.

They occur only when God wills, and only for specific prophetic purposes.

Islam *does* affirm the possibility of karāmāt, extraordinary events granted to especially pious individuals (awliyā'). These are not prophetic miracles and are not meant as public proofs of revelation. They are also understood to be rare and not something anyone is entitled to witness today. Classical Islamic literature includes examples such as the account of ʿUmar ibn al-Khattāb (RA) warning his general Sāriyah from miles away ("Yā Sāriyah, al-jabal!"), and reports of unusual abilities attributed to Shaykh ʿAbd al-Qādir al-Jīlānī in later hagiographical works. Muslims do not equate these with prophecy, but these types of events are closer to what modern parapsychologists classify as *psychic phenomena*. While Muslims may not accept that categorization, there are conceptual similarities in the kinds of unusual events described.

For this reason, the absence of frequent, public miracles today does not require any ad hoc explanation within the Islamic worldview. Prophetic miracles were tied to prophets, and prophets have ended. Karāmāt are by definition rare, limited, and not comparable to the miracles of Prophet Muhammad ﷺ. Therefore, unlike Evolution or the Moon Landing—where the lack of present-day large-scale demonstrations creates tension that invites ad hoc reasoning—the Islamic framework already contains a built-in, coherent explanation for why miracles are not expected in our time.

Another common objection raised is that "only Muslims record Muslim miracles" or "only Christians record Christian miracles." At first glance, this may seem to suggest bias or fabrication. However, when we examine the nature of historical record-keeping more broadly, the argument quickly loses force.

The vast majority of ordinary historical events were recorded primarily—or exclusively—by those within the culture or civilization in which they occurred. For example, many events in Roman history are absent from Persian records, and vice versa. The lack of documentation outside a particular civilization is not surprising, given the limited communication networks, selective historical focus, and logistical constraints of the time. Ancient historians typically recorded events they considered significant for their own society, political interests, or cultural memory, rather than aiming for comprehensive global coverage.

This means that the mere fact that miracles appear mainly in Muslim or Christian sources does not automatically validate or invalidate them. It simply highlights the natural limits of historical documentation. To assess the strength of miracle claims, we should instead examine the evidence for and against the reliability of the particular historical records in question. As I showed earlier in this book, there is an absolutely massive amount of evidence supporting the reliability of Islamic historical records. Those lines of evidence put the islamic miracle reports on a much firmer foundation than a simple appeal to external corroboration.

A common objection to the moon-splitting miracle is that no independent, non-Muslim sources record the event. Critics often argue that such a significant phenomenon should have been observed and documented by contemporaneous civilizations, such as the Byzantine or Sassanian empires. At first glance, this seems to challenge the historicity of the miracle. However, when we consider the historical and practical circumstances, the absence of external records is not only understandable but expected.

First, the event occurred in Arabia at night. Vast regions of the world would not have had nighttime visibility at the same time, and even where it was night, people might not have been looking at the sky. Habits of observation in the 7th century were very different from modern expectations: most people went to sleep earlier, and systematic astronomical recording was not common outside specialized communities. The event likely lasted only a short period—perhaps an hour or two—further limiting the chances of widespread observation.

Second, the historical and political context matters. The emerging Muslim community was in direct conflict with powerful neighboring civilizations, including the Byzantine and Sassanian empires. These states would have had little reason to preserve or transmit reports that might validate a rival religion. Bias, disinterest, or even active suppression could explain the lack of records, without implying that the event did not occur.

Third, extraordinary astronomical phenomena are often underreported outside localized regions, even when they are dramatic. A striking parallel is the supernova of 1054, which was bright enough to be visible during the day for weeks. Yet historical records primarily come from China and Japan, with very few accounts elsewhere. This demonstrates that even widely visible, well-attended phenomena may have sparse documentation depending on geography, cultural priorities, and observational habits. The same reasoning applies to the moon-splitting miracle: the limited number of observers and the short duration make it entirely plausible that few non-Muslim sources noted it.

Taken together, these considerations show that the so-called "ad hoc" explanation for the absence of external records is actually grounded in rational historical and practical reasoning. Unlike the ad hoc claims for evolution—such as the need for millions of years of unobservable change—or for the Moon landing—such as the appeal to financial or technological constraints—these justifications are not speculative. They rely on observable patterns in history and human behavior, demonstrating that the absence of non-Muslim witnesses is neither surprising nor evidence against the miracle.

Conclusion

In examining the common ad hoc objections to Islamic miracles, a clear pattern emerges. Claims that miracles should be occurring today, or that they must be documented by outsiders, misunderstand both the purpose and context of these events. Unlike technological achievements or large-scale historical claims, miracles—especially those of the prophets—are rare, targeted, and meant to confirm divine truth rather than serve as continuous, observable phenomena.

Historical and textual evidence, particularly from the Qur'an and reliable Islamic records, provides strong support for these miracles, while practical considerations—such as geography, duration, and human observational habits—explain why independent verification is limited. When compared to the speculative justifications required to explain phenomena like the Moon landing or macroevolution, the reasoning supporting Islamic miracles is rational, consistent, and grounded in historical reality.

Ultimately, the study of miracles requires applying a consistent standard of inquiry: examining evidence, considering context, and evaluating plausibility. When assessed in this way, the objections based on absence of contemporary observation or external records fail to undermine the credibility of these extraordinary events.

Conclusion

Throughout this book, we have examined claims often presented as settled facts—macro-evolution, the Moon landing, and miracles—through a consistent lens of evidence, observation, and reasoning. In each case, the strength of the claim depends not on popularity, authority, or selective evidence, but on whether it can withstand critical scrutiny.

We have seen that explanations for phenomena not observed today—whether evolutionary transitions, crewed lunar missions, or miracles—often rely on ad hoc reasoning. Yet the nature and justification of these ad hoc explanations vary widely. In the case of miracles, the rarity and purpose of the events, supported by early, reliable historical sources, make the justifications rational and coherent. By contrast, the ad hoc explanations for evolution and the Moon landing require assumptions that are unsupported or inconsistent with observable patterns.

The overarching lesson is clear: claims—extraordinary or otherwise—must be evaluated on the same standards of evidence and reasoning. Consistency, historical context, and critical analysis are essential. Applying these principles reveals which claims are genuinely plausible, which require leaps of faith, and which rely primarily on authority or assertion. This approach empowers readers to discern truth, navigate competing narratives, and uphold intellectual integrity across scientific, historical, and theological domains.

---

[1] D. M. Armstrong, *What Is a Law of Nature?* (Cambridge University Press, 1983).

[2] D. Hume, *An Enquiry Concerning Human Understanding*, Section X ("Of Miracles"); N. Cartwright, *How the Laws of Physics Lie* (Oxford University Press, 1983).

[3] D. Hume, ibid.

[4] N. Cartwright, ibid.

[5] al-Ghazali[1]

[6] D. M. Armstrong, *What Is a Law of Nature?* (Cambridge University Press, 1983).

[7] David Hume, *An Enquiry Concerning Human Understanding*, Section X, in *Hume: Philosophical Works*, ed. T. H. Green and T. H. Grose (Oxford: Oxford University Press, 1896).

[8] Ammar Adil, *The Existence of God and the Irrationality of Atheism* (self-published, 2025).

[9] In the Beginning: An Interview with Olivera Petrovich |[2]

[10] Richard Dawkins, The Blind Watchmaker, 1996, p. 1

[11] Sunan Ibn Majah 63

[12] Sunan An Nasai 4455

[13] Sagan, The Demon Haunted World, 302

[14] Hyman and Honorton, "A Joint Communiqué," 351.

---

1. https://plato.stanford.edu/entries/al-ghazali/?utm_source=chatgpt.com

2. https://jasongoroncy.com/2008/07/28/in-the-beginning-an-interview-with-olivera-petrovich/

[15] Hyman, "Evaluation of Program on Anomalous Mental Phenomena."

[16] Price, "Science and the Supernatural," 359.

[17] Truzzi, "On Some Unfair Practices towards Claims of the Paranormal."

[18] Physicists & Philosophers debunk the Kalam Cosmological Argument featuring Penrose, Hawking, Guth[3]

[19] David Hume, *An Enquiry Concerning Human Understanding* (Oxford University Press, 2000 [1748]), Section X: "Of Miracles

[20] Thomas Reid, *An Inquiry into the Human Mind on the Principles of Common Sense* (Edinburgh: Millar and Kincaid, 1764), ch. 6.

[21] C. A. J. Coady, *Testimony: A Philosophical Study* (Oxford University Press, 1992); Elizabeth Fricker, "Against Gullibility," in *Knowing from Words: Testimony as a Source of Knowledge*, ed. Bimal Krishna Matilal and Arindam Chakrabarti (Springer, 1994);

[22] E. F. Loftus and J. C. Palmer, "Reconstruction of Automobile Destruction," *Journal of Verbal Learning and Verbal Behavior* 13 (1974): 585–589.

[23] G. L. Wells and E. A. Olson, "Eyewitness Testimony," *Annual Review of Psychology* 54 (2003): 277–295.

[24] National Research Council, *Identifying the Culprit: Assessing Eyewitness Identification* (National Academies Press, 2014).

[25] J. T. Wixted, R. Mickes, and R. P. Fisher, "Rethinking the Reliability of Eyewitness Memory," *Perspectives on Psychological Science* 13 (2018): 1–20.

---

3. https://www.youtube.com/watch?v=pGKe6YzHiME

[26] C. Yuille and J. Cutshall, "A Case Study of Eyewitness Memory of a Crime," *Journal of Applied Psychology* 71 (1986): 291–301.

[27] A Visual Analytics based Investigation on the Authorship of the Holy Quran, Halim Sayoud, 2015

[28] Fabricated Hadith Detection: A Novel Matn-Based Approach With Transformer Language Models, K. Gaanoun, M.Alsuhaibani, 2022

[29] The Generalship of Muhammad, Russ Rodgers, p. 16-17

[30] Angelika Neuwirth, *Scripture, Poetry, and the Making of a Community: Reading the Qur'ān as a Literary Text*, 2014

[31] D. S. Margoliouth, *Mohammed and the Rise of Islam*, 1905, p. 123

[32] Holy Qur'an 30:2-5

[33] Holy Qur'an 6:37

[34] Holy Qur'an 10:20

[35] Holy Qur'an 17:59

[36] Matthew 12:38-39

[37] Pfizer accused of Covid profiteering as first-quarter sales hit $26bn[4]

[38] When drug news is no news - PMC[5]

4. https://www.theguardian.com/business/2022/may/03/pfizer-covid-sales-pricing-vaccine-paxlovid-pill?utm_source=chatgpt.com

5. https://pmc.ncbi.nlm.nih.gov/articles/PMC1440632/

[39] The Generalship of Muhammad, Russ Rodgers, p. 16-17

[40] Muhammad Zubayr Siddiqi, The Hadith for Beginners, p. 54

[41] Muhammad Zubayr Siddiqi, The Hadith for Beginners, p. 53

[42] Holy Qur'an 33:9

[43] The Generalship of Muhammad, Russ Rodgers, p. 147-148

[44] Holy Qur'an 54:1–3

[45] Angelika Neuwirth, Cambridge Companion to the Qur'an, p. 100

[46] Benham Sadeghi, Uwe Bergmann, The Codex of a Companion of the Prophet and the Qur'an of the Prophet, 2010

[47] How does the Moon affect the Earth? | Institute of Physics[6]

[48] Natural sleep and its seasonal variations in three pre-industrial societies - PMC[7]

[49] Benchmarks: July 4, 1054: "Birth" of the Crab Nebula[8]

[50] European Historical Evidence Of The Supernova Of Ad 1054 Sky Above Europe On 4th July 1054[9]

---

6. https://www.iop.org/explore-physics/moon/how-does-moon-affect-earth?utm_source=chatgpt.com

7. https://pmc.ncbi.nlm.nih.gov/articles/PMC4720388/?utm_source=chatgpt.com

8. https://www.earthmagazine.org/article/benchmarks-july-4-1054-birth-crab-nebula?utm_source=chatgpt.com

9. https://www.researchgate.net/publication/350371718_European_Historical_Evidence_Of_The_Supernova_Of_Ad_1054_Sky_Above_Europe_On_4th_July_1054

[51] Past-Tense Verbs of Futurity in the Holy Quran: A New Text-World Theory Approach[10]

[52] Hussain 2022 Wisdom in The | PDF[11]

[53] Hussain 2022 Wisdom in The | PDF[12]

[54] Holy Qur'an 17:1

[55] Sahih Bukhari 3584

[56] Sahih Bukhari 3572

[57] Sahih Muslim 1729

[58] David Hume, *An Enquiry Concerning Human Understanding*, ed. Tom L. Beauchamp (Oxford: Oxford University Press, 1999), Section X: "Of Miracles."

[59] C. A. J. Coady, *Testimony: A Philosophical Discussion* (Oxford: Clarendon Press, 1992), pp. 60–78.

[60] John F. Kennedy, "Address at Rice University on the Nation's Space Effort," Sept. 12, 1962, JFK Presidential Library.

[61] Leonard Reiffel et al., *A Study of Lunar Research Flights, Volume I* (U.S. Air Force, Project A119, 1958; declassified 2000).

---

10. https://www.researchgate.net/publication/333040436_Past-Tense_Verbs_of_Futurity_in_the_Holy_Quran_A_New_Text-World_Theory_Approach?utm_source=chatgpt.com
11. https://www.scribd.com/document/658021452/Hussain-2022-Wisdom-in-The?utm_source=chatgpt.com
12. https://www.scribd.com/document/658021452/Hussain-2022-Wisdom-in-The?utm_source=chatgpt.com

[62] Roger D. Launius, "Public Opinion Polls and Perceptions of U.S. Human Spaceflight," *Space Policy* 19, no. 3 (2003): 163–175.

[63] Edward W. Said, *Orientalism* (New York: Pantheon, 1978).

[64] Niall Ferguson, *Empire: How Britain Made the Modern World* (London: Penguin, 2003), ch. 1–2.

[65] Dipesh Chakrabarty, *Provincializing Europe* (Princeton University Press, 2000).

[66] Brian Deer, "How the Case Against the MMR Vaccine Was Fixed," *BMJ* 342 (2011): c5347.

[67] *The Washington Post*, "CDC Vaccine Official Fiona Havers Resigns Amid Data Transparency Dispute," June 2025.

[68] National Center for Science Education, "The Sternberg Peer-Review Controversy," *Reports of the NCSE* 25, no. 5 (2005).

[69] NASA History Office, "Apollo Missions," NASA.gov (accessed 2024); see also Roger Launius, *NASA: A History of the U.S. Civil Space Program* (Krieger, 1994).

[70] J. S. Weiner, *The Piltdown Forgery* (Oxford University Press, 1955).

[71] Robert J. Richards, *The Tragic Sense of Life: Ernst Haeckel and the Struggle over Evolutionary Thought* (University of Chicago Press, 2008).

[72] John S. Mattick, "The Central Role of RNA in Human Evolution and Development," *Trends in Genetics* 17, no. 3 (2001): 139–143.

[73] Christopher B. Coleman, *The Treatise of Lorenzo Valla on the Donation of Constantine* (Yale University Press, 1922).

[74] David King, *The Commissar Vanishes: The Falsification of Photographs and Art in Stalin's Russia* (Metropolitan Books, 1997).

[75] Roger D. Launius, *NASA: A History of the U.S. Civil Space Program* (Krieger, 1994); James Oberg, *Red Star in Orbit* (Random House, 1981).

[76] Qur'an 7:107–108; 26:63 — miracles of Moses, including the staff and parting of the sea.

[77] Qur'an 3:49; 5:110 — miracles of Jesus performed by God's permission.

[78] Ibn Taymiyyah, *Majmūʿ al-Fatāwā*, vol. 11, 307–309 — distinguishing karāmāt from acts of sorcery or jinn.

[79] Stephen Jay Gould, "Punctuated Equilibrium," *Nature* 366, no. 6452 (1993): 223–227.

[80] Roger D. Launius, *Apollo's Legacy: Perspectives on the Moon Landings* (Smithsonian Books, 2019).

[81] ENCODE Project Consortium, "An Integrated Encyclopedia of DNA Elements in the Human Genome," *Nature* 489, no. 7414 (2012): 57–74.

[82] Laurin, M. "Appendix function and evolution," *Biological Reviews* 96, no. 2 (2021): 545–559.

[83] American Psychiatric Association, *Diagnostic and Statistical Manual of Mental Disorders*, 5th ed. (Washington, D.C.: APA, 2013).

[84] Bartholomew, R.E. & Wessely, S., "Protean nature of mass psychogenic illness: From possessed nuns to chemical and biological terrorism fears," *British Journal of Psychiatry* 180 (2002): 300–306.

[85] Rankin, K.M., "The Tanganyika Laughter Epidemic: A Study of Mass Psychogenic Illness," *Psychiatric Quarterly* 74, no. 2 (2003): 173–183.

[86] Smallman-Raynor, M. & Cliff, A., "Epidemics of mass hysteria and collective stress: A study of social transmission," *Social Science & Medicine* 45, no. 10 (1997): 1489–1499.

[87] John Walton, *Ancient Near Eastern Thought and the Old Testament* (Grand Rapids: Baker Academic, 2006), pp. 34–36, discussing typical divergences in eyewitness accounts of historical events.

[88] Craig S. Keener, *Miracles: The Credibility of Historical Accounts* (Grand Rapids: Baker Academic, 2011), pp. 45–60, on how variations in accounts of miracles are consistent with normal historical reporting.

[89] *Expelled: No Intelligence Allowed,* directed by Nathan Frankowski (Premise Media, 2008).

[90] E.V.R. Kojonen, The Intelligent Design Debate and the Temptation of Scientism, 2015, p. 55

[91] Charles Darwin, *On the Origin of Species* (London: John Murray, 1859), 466.

[92] Simon Conway Morris, *Life's Solution: Inevitable Humans in a Lonely Universe* (Cambridge: Cambridge University Press, 2003), 82.

[93] Stephen Jay Gould, *The Structure of Evolutionary Theory* (Cambridge, MA: Belknap Press, 2002), 1433.

[94] William A. Dembski, *Intelligent Design* (Downers Grove, IL: InterVarsity Press, 1999), 256.

[95] Robert T. Pennock, *Tower of Babel: The Evidence Against the New Creationism* (Cambridge, MA: MIT Press, 1999), 15.

[96] E.V.R. Kojonen, The Intelligent Design Debate and the Temptation of Scientism, 2015, p. 56

[97] Fisher, The Genetical Theory of Natural Selection, 44

[98] McDonald, "The Molecular Basis of Adaptation," 93

[99] McDonald, "The Molecular Basis of Adaptation," 93

[100] Elliott Sober, Evidence and Evolution: The logic behind the science, 2008, p. 264, emphasis added

[101] Elliott Sober, Evidence and Evolution: The logic behind the science, 2008, p. 265

[102] Richard Dawkins, The God delusion, page 169-170

[103] Physicists & Philosophers debunk The Fine Tuning Argument[13]

[104] Convergent evolution explained with 13 examples | Natural History Museum[14]

[105] Jerry A. Coyne, Why Evolution Is True, 2009, p. 71-72

[106] Gould, S.J., Is a new and general theory of evolution emerging? Paleobiology 6:119–130 (p.127), 1980.

[107] Gould, S.J., Evolution' erratic pace. *Natural History* 86(5):14, 1977.

---

13. https://www.youtube.com/watch?v=jJ-fj3lqJ6M

14. https://www.nhm.ac.uk/discover/convergent-evolution.html

[108] Niles Eldredge, Reinventing Darwin: The Great Evolutionary Debate (New York: John Wiley & Sons, 1995), 95.

[109] Stephen Jay Gould: Did He Bring Paleontology to the "High Table"?[15]

[110] Absolute measures of the completeness of the fossil record | Nature[16]

[111] Stephen Jay Gould, Punctuated Equilibrium, 2007, p. 19

[112] Schopf, Editorial Introduction to Eldredge and Gould, "Punctuated Equilibria: An Alternative to Phyletic Gradualism"

[113] Neo-darwinism still haunts evolutionary theory: A modern perspective on - PMC[17]

[114] Charlesworth, Lande, and Slatkin, "A Neo-Darwinian Commentary on Macroevolution," 493

[115] Valentine and Erwin, "Interpreting Great Developmental Experiments," 96

[116] Shoaib Ahmed Malik, Islam and Evolution: Al-Ghazālī and the Modern Evolutionary Paradigm, 2021, p. 38

[117] Vestigial organs: Remnants of evolution | New Scientist[18]

[118] An integrated encyclopedia of DNA elements in the human genome | Nature[19]

---

15. https://www.researchgate.net/publication/43225497_Stephen_Jay_Gould_Did_He_Bring_Paleontology_to_the_High_Table

16. https://www.nature.com/articles/18872

17. https://www.ncbi.nlm.nih.gov/pmc/articles/PMC8979413/

18. https://www.newscientist.com/article/mg19826562-100-vestigial-organs-remnants-of-evolution/

19. https://www.nature.com/articles/nature11247

[119] Laura Poliseno, "Pseudogenes: Newly Discovered Players in Human Cancer," Science Signaling, 5 (242) (September 18, 2012).

[120] Elliott Sober, Evidence and Evolution: The logic behind the science, 2008, p. 129

[121] Todd, S.C., correspondence to Nature 401(6752):423, 30 Sept. 1999

[122] Dionysius of Tel-Mahre, The Seventh Century in the West-Syrian Chronicles, Liverpool, 1993

[123] Houston, we have a problem: original moon walk footage erased | Apollo 11 | The Guardian[20]

[124] NASA Discarded Hardware From Apollo Missions to the Moon - Newsweek[21]

[125] Absolute measures of the completeness of the fossil record | Nature[22]

[126] The completeness of the fossil record - Benton - 2009 - Significance - Wiley Online Library[23]

[127] Quality of the fossil record through time | Nature[24]

[128] Sampling, taxonomic description, and our evolving knowledge of morphological diversity | Paleobiology | Cambridge Core[25]

---

20. https://www.theguardian.com/science/2009/jul/16/moon-landing-tapes-erased?utm_source=chatgpt.com

21. https://www.newsweek.com/nasa-discarded-hardware-apollo-missions-moon-1739500?utm_source=chatgpt.com

22. https://www.nature.com/articles/18872

23. https://rss.onlinelibrary.wiley.com/doi/full/10.1111/j.1740-9713.2009.00374.x

24. https://www.nature.com/articles/35000558

25. https://www.cambridge.org/core/journals/paleobiology/article/abs/sampling-taxonomic-description-and-our-evolving-knowledge-of-morphological-diversity/5D44EF179954D44ED12A77D18D7DAA22

[129] Structure, not Bias | Journal of Paleontology | Cambridge Core[26]

[130] The quality of the fossil record across higher taxa: compositional fidelity of phyla and classes in benthic marine associations - PMC[27]

[131] George McGhee, Convergent Evolution: Limited Forms Most Beautiful, 2011, p. 39-40

[132] Simon Conway Morris, Life's Solution: Inevitable humans in a lonely universe, 2003, p. 283-284

26. https://www.cambridge.org/core/journals/journal-of-paleontology/article/structure-not-bias/309B2D6D87176F77774D07A67B1E7A64

27. https://www.ncbi.nlm.nih.gov/pmc/articles/PMC10348303/

Also by Ammar Adil

Sunni Hadiths and Sirah: Proofs of Historical Reliability
The Existence of God and the Irrationality of Atheism
Proofs for Islam: Prophecies of Prophet Muhammad ﷺ
Hadiths Sunnites et Sirah : Preuves de Fiabilité Historique
Evolution and Design: Logic and Evidence
Evrim ve Tasarım: Mantık ve Kanıt
Sunnitische Hadithe und Sirah: Beweise historischer Zuverlässigkeit
Sünni Hadisler ve Siyer: Tarihsel Güvenilirliğin Delilleri
From Skepticism to Certainty: The Doubting Muslim's Guide
Évolution et Conception : Logique et Preuves
Evolution und Design: Logik und Beweise
Du scepticisme à la certitude : le guide du musulman qui doute
Vom Skeptizismus zur Gewissheit: Der Leitfaden für zweifelnde Muslime
Beyond Scientific Miracles: Islam, Science, and the Search for Truth
L'Existence de Dieu et l'irrationalité de l'athéisme
Tanrı'nın Varlığı ve Ateizmin Mantıksızlığı
Belief, Evidence, and the Unseen: Miracles, Evolution, and the Moon Landing

About the Author

I am an independent researcher in topics related to existence of God and Islam. I done years of research on topics such as Existence of God, Evolution, Intelligent Design, Proofs for Islam being from God, Historical Jesus, Quran Preservation, Reliability of Hadiths and Sirah etc. My debates and videos are uploaded on https://youtube.com/@refutingorientalists?si=2Q8ZzGfsupml-IeQ and you can contact me on my twitter account at https://twitter.com/qadissiyah636

About the Publisher